David Esdaile

Contributions to Natural History

Chiefly in Relation to the Food of the People

David Esdaile

Contributions to Natural History

Chiefly in Relation to the Food of the People

ISBN/EAN: 9783337026462

Printed in Europe, USA, Canada, Australia, Japan

Cover: Foto ©berggeist007 / pixelio.de

More available books at **www.hansebooks.com**

CONTRIBUTIONS

TO

NATURAL HISTORY

CHIEFLY IN RELATION TO

THE FOOD OF THE PEOPLE

BY

A RURAL D.D.

WILLIAM BLACKWOOD AND SONS
EDINBURGH AND LONDON
MDCCCLXV

PREFACE.

LIVING in the country, and possessing a taste for Natural History, it is not surprising that I should have been led to write about surrounding objects.

Contributions to various periodicals, especially to 'The Quarterly Journal of Agriculture,' have thus in the course of time become so numerous that it has been thought desirable to publish a selection from them in the present form; revised, that is, and—it is hoped—improved by being curtailed or added to as seemed to be necessary.

The alimentary virtues of Horse-Flesh and Fungi are discussed with a half-earnest levity: I use the language of earnest conviction when directing public attention to the great storehouse within which the Universal Parent has laid up exhaustless supplies of food. We may be squeamish as to the eating of *cheval* soup and other dainty preparations from the flesh of the horse; fear may hinder us feasting on Fungi; but with the sea as our fish-pond, and pisciculture capable of endless development alike in salt-water and in fresh, it is marvellous that multitudes should be pining with hunger in a land like ours, literally set in an ocean of plenty.

The interests of the public economy, as well as of large sections of our maritime communities, are so deeply implicated in oyster and mussel culture, and in the judicious prosecution of fishing for herring and salmon, that no regret is felt for the time expended in making known the resources of pisciculture at home and abroad.

D. E.

September 1865.

CONTENTS.

	PAGE
HIPPOPHAGY; OR, SHOULD WE EAT OUR HORSES?	1
THE PROGRESS OF HIPPOPHAGY: A PLEA FOR EATING HORSE-FLESH,	17
MYCOPHAGY; OR, SHOULD WE EAT FUNGUSES?	31
LIFE AND HISTORY OF A SALMON,	57
SALMON AND PISCICULTURE,	80
SCOTCH SALMON AND SCOTCH LAW,	88
THE SALMON RIVERS OF ENGLAND AND WALES,	107
SALMON—BRITISH AND COLONIAL,	122
SALMON-REARING AT STORMONTFIELD, AND FISH-CULTURE,	140
SOMETHING MORE ABOUT THE HATCHING OF FISH,	154
THE HERRING,	168
POPULAR WEATHER PROGNOSTICS,	194
HIRUDICULTURE (LEECH-CULTURE),	215
MARITIME PISCICULTURE—LAGOON OF COMACCHIO,	226
MARITIME PISCICULTURE—OYSTER-CULTURE,	242

	PAGE
MARITIME PISCICULTURE—MUSSEL-CULTURE,	260
FISH DIET, AND ITS EFFECTS ON THE HUMAN CONSTITUTION,	271
PEARLS AND PEARL-CULTURE,	280
HORSES—ANCIENT AND MODERN,	288
THE ARAB HORSE OF AFRICA,	312
ACCLIMATISATION SOCIETIES,	337

CONTRIBUTIONS

TO

NATURAL HISTORY.

―◆―

HIPPOPHAGY;

OR, SHOULD WE EAT OUR HORSES?*

Doctor Johnson's beautiful story of 'Rasselas' has made everybody acquainted with the Abyssinian prince who, satiated with the good things of this life, offered a noble reward to the inventor of a new pleasure. The pleasure which the learned, eloquent, and most benevolent Professor of Zoology to the Faculty of Sciences at Paris proposes for our acceptance is certainly not new; for many a comfortable meal of nutritious horse-flesh—ay, of ass, mule, and zebra flesh—has been made of old time, and by many nations further advanced in acquaintance with alimentary substances than we con-

---

* 'Lettres sur les Substances Alimentaires, et particulièrement sur la Viande de Cheval.' Par M. Isidore Geoffroy Saint-Hilaire, Membre de l'Institut, Professeur de Zoologie à la Faculté des Sciences de Paris, &c. Victor Masson; Paris.

ceited moderns are to this day. But the varied learning, the literary skill, the benevolent enthusiasm, the noble contempt for that peculiar cachination for which the equine race gets credit, although only from man do we hear what *he* chooses to call a *horse-laugh*,—these are displayed by our learned Professor with a profusion by which we have been most agreeably surprised. We took up his book with no great expectation of being either gratified with a literary treat, or converted to the faith of the, as yet, small and derided sect of the *Hippophagi*. And yet here are we seriously about to tell the public that M. Saint-Hilaire's Letters are most interesting, and that to laugh at the idea of eating horse-flesh is great folly in these our days, when so many are so unable to answer the daily-recurring question, "What shall we eat?" and when so many must receive from charity their daily bread.

Wise and intellectual folk may turn away in affected indifference to flesh-pots and creature-comforts. As for ourselves, we are not ashamed to confess that, in Count Rumford's Essays, we have read with great delectation a most appetite-raising chapter, entitled "Of the Pleasure of Eating, and of the Means that may be employed for increasing it." Moreover, knowing how intimately interblended are the moral and the mental with the physical elements of our wondrous being, we subscribe, with a moderate allowance due to the *mot* of a wit, to the aphorism of a contemporary Frenchman: "Tell me what you *eat*, and I'll tell you what you *are*."* *Mens sana in corpore sano* being the perfectibility of human nature in this life, a treatise on the best mode of keeping soul and body together—on alimentary substances, in short—is deserving of the most careful study, because involving principles for the evolution and application of which there is required the highest philosophy. Starve a man, or maintain him by a minimum expenditure of nutrition,

* "Que je sache ce que tu manges, et je saurai qui tu es."

and what a miserable creature do you make him! Stunted in his physical organisation, his mental powers share in the resulting debility; and how readily our moral perceptions sympathise with our bodily condition, we all of us occasionally know, when disordered health tries our temper and obscures our judgment. We therefore assent to this characteristic declaration in one of the earliest productions of the benevolent Dr Chalmers: "Let it be remembered that Philosophy is never more usefully, and never more honourably, directed, than when multiplying the stores of human comfort; than when enlightening the humblest departments of industry; than when she descends to the walks of business, to the dark and dismal receptacles of misery; to the hospitals of disease, to the putrid houses of our great cities, where Poverty sits in lonely and ragged wretchedness, agonised with pain, faint with hunger, and shivering in a frail and unsheltered tenement. Count Rumford deserves the gratitude of mankind." This panegyric on Count Rumford, better known to the scientific world under his English title of Sir Benjamin Thomson, is abundantly merited. His Essays are full of important experiments in nutrition, and from them are derived almost all our so-called novelties in modes of heating and ventilation, as well as in the construction of cooking utensils of every kind. From them the philanthropist will learn with delight how, on New-year's Day, the philosophic Count captured all the swarming beggars of Munich, introduced them into a military workhouse, and soon made them fat, happy, and industrious. From them also the careful housewife will learn with surprise how truly says the poet,

"Man needs but little here below;"

for the Bavarian soldier, who is very fond of eating, and whose situation is as comfortable as that of any soldier in Europe, lives, we are told, on twopence a-day; so skilled is he in the science of cookery.

If our old friend Count Rumford deserves to live in the public memory as a benefactor to the human race, because of his benevolent ingenuity in turning to the best account the usual articles of food, M. Saint-Hilaire claims a niche in the temple of fame on account of his eloquent and scientific demonstration of the excellent qualities of a species of food the prejudice against the use of which, in Europe at least, is inveterate, and almost universal. The human stomach, at least when hungry, is not apt to be sentimental; but it is astonishingly apt to be squeamish and whimsical. Hence the strenuous efforts of his Government have only partially induced the white-snail-eating Austrian to partake of horse-flesh; and if Louis Napoleon were to declare it his imperial pleasure that Paris should consume a large proportion of this really excellent food, what a dangerous commotion would there be among the frog-eaters on the banks of the Seine!

M. Saint-Hilaire is well aware of the inveteracy of the prejudice which obstructs his philanthropy; for nine years he exposed its folly in the presence of the enlightened audiences which yearly listened with delight to his lectures in the Museum of Natural History; and, so anxious was he to give it the *coup-de-grace*, that for several months he suspended the publication of his 'General Natural History,' in order that he might publish the thoroughly practical work on which we are now commenting. "May it," he exclaims, "be the death-blow to that silly prejudice against which I have been contending for nine years, and against which I shall contend so long as I witness the deplorable spectacle of millions of Frenchmen deprived of animal food; eating it six times, twice, *once* a-year! and in presence of this misery millions of pounds of good meat given over to industry every month for secondary purposes, abandoned to swine and dogs, or even cast into the dunghill."

To know that there are many as miserable as themselves is a kind of comfort. We therefore compassionately remind these distressed Frenchmen, that though to be omnivorous is one of the distinctions betwixt man and the brutes, yet, in fact, only a small proportion of the human race is actually carnivorous; and that for more than sixteen hundred years the entire population of the earth was restricted to a vegetable diet; for not till after the Flood did the Almighty say to Noah and his sons, "Every moving thing that liveth shall be meat for you; even as the green herb have I given you all things."—(Gen. ix. 37.) As an additional crumb of comfort for non-flesh-eating Frenchmen, we add, that the idea of "John Bull" living on animal food is unhappily a fiction. "If 'John Bull' means two-thirds of the population, 'John Bull' is living on vegetable diet, and not more than one-third of him is nourished by meat."* Speaking of the Dorsetshire peasantry, Mr Thornton, in his work on Over-Population, observes: "As for meat, most of them would not know its taste, if once or twice in the course of their lives—on the squire's having a son and heir born to him, or on the young gentleman's coming of age—they were not regaled with a dinner of what the newspapers call 'Old English Fare.' Some of them contrive to have a little bacon—in the proportion, it seems, of a half-pound a-week to a dozen persons; but they more commonly use fat, to give the potatoes a relish; and, as one of them told Mr Austin, they don't always go without cheese." In Scotland, the mass of the people are compulsory vegetarians, oatmeal and milk being the national fare. Pigs are kept by the Scottish peasantry very generally, but often they are not eaten by their families, but sold, owing to the high price given for pork by *krämers*;† that is, men who go through the country buying pigs, butter, and eggs

---

* Symon's 'Arts and Artizans.'

† *Krämer* is in Germany the term for a petty huxter, or small merchant. It is curious to find the word similarly applied in Forfarshire.

for London and other great cities. Ireland is notoriously the land of pigs and potatoes, but unfortunately the Irish peasant does not eat but sells his pig—" the gentleman that pays the rint."

If the cold of these northern regions did not sharpen the appetite, and demand the use of fatty substances for the comfortable nutrition of human beings, the people of the British Isles would have little reason to complain because of being generally restricted to vegetable fare, small acquaintance with flesh-pots being the lot of the people almost everywhere; a mere fraction of the human race being carnivorous, though furnished with means of masticating and assimilating animal food. But with our climate, and with the active labours required from most of us, the liberal use of animal food is indispensable for the development and maintenance of our bodily vigour. We lately read with amazement of the vegetable fare solely employed by a singularly lusty English blacksmith—a teetotaller, moreover. But how long will his strength endure? We beseech him to meditate on the fate of M. le Docteur Stard, who, trying a philosophical experiment, died while flattering himself that he was only weakened by a vegetable diet. Let him also meditate on this novel illustration of the saying that "hunger will tame a lion." A lion, fed for many years on milk-soup! was presented in 1855 to the Menagerie of the Museum of Natural History, Paris. The poor brute was as quiet as a sheep, and so debilitated as to be *in extremis*. An instant change of diet was resolved upon. In a month horse-flesh restored his natural ferocity, and now he is magnificent! In constitution, a brose-fed Scottish ploughman is on the verge of old age in his fiftieth year. We invite attention to this fact, and we trust those interested in the public *hygiène* will ponder well the distressing statement that in some parts of England scanty nutrition is unfitting the labourer for toil, and, by weakening his thews and sinews, incapacitating him for serving his country, either in peace or in war; for guiding the plough, or wielding sword and bayonet. The recruiting officers in the south-eastern counties complain that in size and strength the peasantry are inferior to those of the north and north-midland counties, and farmers observe the same variation in the amount of labour which they can obtain from their men.* This deteriorated race gives birth to a yet more enfeebled offspring; so that the desire to arrest this degeneration of the physical constitution of the British people should make us give willing ear to such a man as M. Saint-Hilaire, when propounding his views on a matter of such importance as that regarding the means of public nutrition.

He discusses, in the first place, the normal laws of nutrition established by Liebig and other eminent chemists, who demonstrate that flesh-eating animals are in general stronger than the herbivorous on which they prey, and that no other substance equals animal food in the production of flesh, and in the reparation of muscular energy expended in labour. Hence the necessity of providing it for the labouring classes everywhere, and especially in northern climates and in great cities. Is it sufficiently supplied? Evidently not. Adopting 83 kilogrammes a-year as the normal rate of nutrition necessary for the comfortable existence of human beings, M. Saint-Hilaire demonstrates that this is so far from being attained in France, that to arrive at it demands the production of three and a-half times more animal food than France actually produces. Instead of 83 kilogrammes, Frenchmen, on the average, only consume 28 kilogrammes of animal food per annum. " The difference between the normal and the actual consumption is enormous," exclaims the Professor, "the deficit immense!" We need hardly inform our readers that he proposes to supply this deficit by the consumption of horse-flesh, "a reserve for which we need not cross the sea, nor even the frontier, which is always at hand,

* Thornton on Over-Population.

and the benefit of which we may have to-morrow if we please to will it to-day."

Whether M. Saint-Hilaire ever heard of the well-known recipe for hare-soup, beginning with, "First catch your hare," we know not. Certain it is that he, very considerately, first informs us where we may find the horses on which he invites us to feed. With most praiseworthy precision he states the relative proportions of the various animal products consumable in France; from which it results that the supply of horse-flesh is equal to one-fourth of the animal food consumed at present. "Singular social anomaly, the long endurance of which will one day excite amazement! There are millions of Frenchmen who never eat flesh, while every month millions of pounds of good meat are applied to very secondary uses, or even thrown into the dunghill. Behold what science herself has authorised up to this day, at least by her silence! as if even she were afraid to oppose a popular prejudice, and to open her hand and spread abroad useful truths which she had in her possession."

"After the question of quantity comes that of quality,"—so begins part second of these interesting Letters. This is the point where our Professor most needs his rhetorical skill, and appeals to facts and testimonies of the enlightened few who, in Europe, have tried Hippophagy. His appeal to our palate is so irresistible that, being about to sit down to dinner, we are heartily sorry at not having the prospect of seeing at our table *vol-au-vents d'amourette* from the spinal marrow of a horse; nor horse-soup, nor horse-pie *à la mode;* nor a roast of horse-chine,—all of which, we know, were lately received with "explosions de satisfaction" by a party of Parisian *Hippophagi.*

Reasoning from analogy, we should expect little difference in the nutritious qualities of the flesh of the horse and the ox. Like our best butcher-meat animals, the horse is herbivorous. The only notable difference is the greater amount in the flesh of the horse of the principle of *kreatine*,—that nitrogenous substance discovered in 1833 in beef-soup by Chevreul, who has found it in all the vertebrate animals, and which, according to Liebig, plays an important part in almost all vital actions. This excess of *kreatine* should add to the alimentary value of horse-flesh. Theory and experiment are on this point in happy union; and M. Saint-Hilaire quotes varied and most competent testimony showing its superiority. We can only refer to the experience of the illustrious surgeon Larrey, who thus sums up his long and distinguished practice in camps and hospitals:—

"I have very often, *and with the greatest success*, given horse-flesh to the soldiers and the wounded of our armies. In several of our campaigns on the Rhine, in Catalonia, and the maritime Alps, I caused it, under various circumstances, to be given to our soldiers; but, above all, we found the very great benefit of this meat during the siege of Alexandria in Egypt. Not only did it save the lives of the troops defending that city; it *powerfully contributed to the cure and invigoration of the numerous sick and wounded in our hospitals, and likewise aided in the removal of a scorbutic epidemic which seized the whole army.* There was a regular daily distribution of this meat; and most fortunately the number of horses was sufficient to bring the army up to the time of the capitulation. These animals, of the Arab breed, were extremely thin, owing to the scarcity of fodder, but they were generally young. In order to overcome the prejudice of the soldiers, I was the first to kill my horses and eat this food.

"After the battle of Esslingen, shut up in the island of Lobau, with the greater part of the French army and about six thousand wounded, I caused soup to be made of the numerous horses scattered over the island, and belonging to the generals and superior officers. The breastplates of those dismounted, and of the wounded, served as coppers for cooking this meat; and instead of salt, of which we were destitute, it was seasoned with gunpowder. I only had the trouble of pouring the soup from one breastplate into another through a linen cloth, and of then allowing it to clarify by rest. Marshal Massena, commander-in-chief of the troops, was right glad to share in my repast, and was very well pleased with it. Experience thus demonstrates that horse-flesh is most proper nourishment for man."

Oh that this had been remembered in the Crimea! We should never have heard of the sufferings of the wretched horses which crunched each other's tails; and many a sick and wounded man might have received that nourishment for want of which he pined and died on the bleak plateau around Sebastopol. In the debate on the Crimean Commission, in the House of Commons, General Peel, speaking of the want of forces for the making of a road from Balaklava to Sebastopol, observed that "it had originated from the impossibility of finding forage for more than a certain number of animals in the Crimea, and that number was already exceeded by the horses of the cavalry, the artillery, and others. The common-sense view would have been to reduce the number of horses to the power of feeding them. The proper course would have been to have re-embarked a portion of them." Not a hoof of them, say we. They should have been slaughtered and eaten, instead of being permitted to die by inches, and their carcasses to diffuse the odour of the most fetid corruption in the vicinity of over-wrought, under-fed, sick and wounded men. What was excellent food for Frenchmen would have been equally good for British troops; and if Massena and Larrey thought horse-flesh dainty fare when seasoned with gunpowder, we are very decidedly of opinion that General Peel would have no reason to complain were he doomed for a few months to a dietary such as nerved the French defenders of Alexandria, and the isolated troops in the isle of Lobau. The conclusion of the distinguished Parent-Duchâtelet, the highest authority in all that affects the public health, is in these words :—" This kind of food was very good and much sought after in ancient times. It has not changed its nature, and it is as suitable for modern stomachs as it was for those of our ancestors; for stomachs strong and healthy, but also for the sick and wounded, whose strength it restores, and whose convalescence it confirms. And it is not necessary that the animals should be fat, or that they should have never suffered, as some may suppose; for beneficial effects may be obtained from horses extremely emaciated by famine." So that we are not amusing our readers with the unprofitable talk of a mere *littérateur*, but communicating the knowledge of facts, by the application of which the wants of the poor may be supplied, the sick and debilitated restored and invigorated, the health of armies maintained, great battles won, and prolonged sieges endured.

But " what is the *gout* of horse-flesh?" Have patience, courteous readers, and we shall make your teeth water! Horse-dinners *vulgo* (*banquets hippophagiques*, in Parisian phrase) have been quite the vogue of late in Paris, Kœnigsbaden, Wirtemberg, Weimar, Munich, Vienna, Dresden, and many other places; many of those in Germany being by public subscriptions. Those in Paris have been especially *recherchés*, and attended by such a variety of distinguished guests as to elevate them to the rank of scientific experiments. Though in possession of the culinary triumphs of the German *Hippophagi*, we shall spare our readers the recital, and give them instead an account of a Parisian "horse-dinner," from the graceful pen of the witty *M. le Docteur Amédée Latour*, who takes care to inform us that it was not written on his rising from table, but twenty-four hours after, when, he solemnly depones, he was suffering not the least digestive remorse. " The experiment begins. M. Renault has most intelligently made the arrangements. Side by side are the subjects to be experimented on—the matters to be compared.

" Horse-soup—Beef-soup.
Horse-boil—Beef-boil.
Roast-horse—Roast-beef.

" The same quantity, the same sort—judge and compare—nothing better.

" *Horse-soup*—general astonishment! It's perfect! it's excellent! it's feeding! it's like venison! it's aromatic! it's rich-tasted! it's a first-rate and admirable soup!

" The *beef-soup* is good, but comparatively inferior, of less marked *gout*, less flavoured, less tasty. The jury unanimously find that the horse yields soup of superior quality, that it is impossible to distinguish the taste of it from that of the richest beef-soups, and that persons not warned could not find out the difference. The same colour, the same clearness.

" *Boiled-horse*—the flesh is browner than beef; it is drier, and less resisting under the teeth; no particular taste; it is the taste of boiled beef, but not of the first class. I have eaten better beef, but also much worse. Upon the whole, it is very eatable; poor people who buy inferior beef, or cow-beef, would find a very sensible difference in favour of boiled-horse.

" *Roast-horse.*—It is the chine of the animal slightly salted and highly spiced. An explosion of satisfaction! *Nothing finer, more delicate, or more tender.* The fillet of venison, whose aroma it recalls, is not its superior. *It's perfect in all points.*

" Summing up: *Soup—superior.*
Boil—good, and very eatable.
Roast—exquisite.

" Is not this a very interesting experiment?"

Truly yes, say we. And here is another, rather

comical. M. Saint-Hilaire was President of the Society of Acclimatation. Having invited a member of this society to taste of *a kind of meat undoubtedly new to him*, the learned doctor thought his opinion was sought for in regard to some rare and newly-introduced animal; and so, after having duly tasted it, he gave it thus: "In my opinion it is of *the utmost importance* to acclimatise this animal." It was horse-flesh!

These things being so, how comes it that there is such a prejudice against such a valuable article of food? Our Professor hunts down this prejudice wherever displayed. He first falls foul of certain Chinese doctors who have interdicted the use of horse-flesh in their much-admired production, *Chi-wou-pen-thsao-hoei-tsouan*. And certainly he does make mince-meat of these poor Chinese, who, in the plenitude of their wisdom, declare that to eat of a *white* horse with a *black* foot, or of a *white* horse with a *black* head, will make a man mad. These worthies also teach that to hang up a monkey in a stable is an infallible preventive of all horse-diseases.

In reference to European prejudice, M. Saint-Hilaire remarks: "One cannot directly attack, as I have attacked, an old idea without encountering the warmest opposition, any more than you can pull up a deeply-rooted tree without vigorous efforts." And so he rides his hobby with a firm seat, and is as little daunted by the folly of fools as by the sneers of witlings, or the objections of reasonable people. He discusses with much dignity and good temper the objections of his learned friends, MM. Valenciennes and Milne Edwards, and proves that they are in error when supposing that only the flesh of young horses is good, inasmuch as most satisfactory meals have been made upon animals from seventeen up to twenty-five years old. To certain objectors, to whom he declines to apply the epithet *savants*, who allege that the public sale of horse-flesh would excite among those using it a feeling of ill-will against the consumers of superior butcher-meat—such as beef, veal, and mutton—he says: "Why not on this principle renounce also butcher-meat? Those who only eat beef might be jealous of those who eat poultry and game." As to the rise in the price of horse-flesh, and the consequent limitation in the use of it when its consumption shall become general, it is argued that it will be long before this happens; and that, when it does happen, it will be a public benefit, by reducing the price of butcher-meat. One fourteenth of an addition to the meat consumed at present will infallibly arrest that rise in the price of meat, of which in France there are such complaints. As to the jokes of a portion of the press, our philosopher laughs when they are witty; when heavy he heeds them not. *Telum imbelle sine ictu.* A certain religious journal is afraid that eating of horses must end in eating of men! "When all the horses are slaughtered, men must eat one another." Our Professor looks very grave on being thus accused of preaching *Anthropophagy!* As to the sentimentalists who declaim against slaying and devouring an animal which is our friend, companion, and servant in our labours, pastimes, and wars, their whining is disposed of by the fact that the horse-eating movement is mainly supported by the societies instituted for the prevention of cruelty to animals. It is undoubtedly more merciful to fatten an old horse, and then eat him, than to work him till he is a moving skeleton, a mass of sores, a sight so piteous as to call forth the indignation of every rightly-constituted mind. Those savage abusers of a noble animal, who torture and starve the horse, may not listen to your humane interpositions in its behalf; but they will hear you when demonstrating that it is their interest to send their horses to the knackers in tolerable condition; because as they (*i. e.*, the horses) are to be eaten, it is manifest that while a skin full of bones may be worth five shillings only, a much higher price will be given for a horse tolerably plump.

In conclusion, we protest against the senseless waste of horse-flesh which has hitherto prevailed; we invite public trial of its qualities; and in order that our readers may have some idea of the ability of the Scottish farmer to meet the demands of those we convert to *Hippophagy*, we beg to add that, according to the agricultural statistics for 1856, there were in Scotland very nearly 180,000 horses. And M'Queen's statistics inform us that in the British empire there are 2,250,000 horses, valued at £67,000,000. A very considerable proportion of these should be eaten. Malformation, incipient disease, accidental injuries, render it more profitable to kill than to rear, at great expense, animals incapacitated for labour. Send an injured or enfeebled horse to grass, and in four months he will be fat and fit for the table. In answer, therefore, to our next query, "Should we eat our horses?" we reply, "Under certain circumstances, undoubtedly;" and in no case ought we to bury the flesh of the horse, when we have the strongest reasons for regarding it as alike nutritious and palatable. Owing to the great value of the horse, the use of its flesh, as an article of human food, must in this country be limited. Let us, however, avail ourselves of it so far as practicable; and in order that it may no longer be wasted, let us dismiss the silly prejudice which causes even a half-starved labourer to exhibit irrepressible disgust when exhorted to partake of a kind of food occasionally within his reach, and known to possess in abundance those elements of nutrition so scantily supplied by the common fare of the working classes.

We invite our readers to test its value, as we have recently done, in a spirit of philosophical inquiry, and, we solemnly declare, without the slightest digestive remorse, though to the horror of our cook, a privileged old servant, who rules the roast, and us too sometimes! Misled by her imagination, and not following her nose assuredly, she declared that she perceived "a fearfu' smell before the stuff cam' near the door." After we had eaten delicious soup and excellent stew, prepared by her reluctant hands, the worthy woman thought she should pray for us. "The Lord be wi' the maister! After eatin' o' sic an onnat'ral brute, he's shure to tak' the worst kind o' jaundice, for it's aye ta'en wi' an awfu' scunner."*

We are happy to announce that the jaundice was all in her eye, and that we are enjoying peace of mind and body after this first, but we hope not last, trial of hippophagy.

If any exclaim, "Horrid man! to advise the eating of horse-flesh,"—quietly, good friends, if you please! Have *you* not eaten of it disguised as Bologna sausage, or "Russian ox-tongues," which, though of undeniable merit, are well known to be those of horses?

* *Anglicè*, intense disgust.

THE PROGRESS OF HIPPOPHAGY:

A PLEA FOR EATING HORSE-FLESH.

IN the year 1857 we were converted to the derided faith of the hippophagi. With the zeal of new converts we gave, to all who would hear us, many good reasons for believing that it is salubrious and appetising to eat of horse-flesh, prepared in the varied modes with which we are familiar in regard to the kinds of meat common at our tables.

Unfortunately we were not practically hippophagous; we had never tasted horse-flesh. A lady roasted us unmercifully on our presumption in prescribing for the public stomach an alimentary substance which had never entered our own. Her ceaseless jibe was "Eat a horse, and I'll believe you." We could not parry this home-thrust. Being in London, we were seriously meditating an excursion to the Continent in order there to join in a *banquet hippophagique*, and so acquire that experimental acquaintance with horse-flesh for lack of which the lady laughed alike at our dietetic philosophy and our philanthropic protestations of desire to augment the food of the people. This pilgrimage was fortunately unnecessary. We were lucky enough to be able to "eat a horse" at home, with the most happy results.

Though much laughed at, we have found imitators, disposed so to become, we daresay, by having to pay

10d. per pound for beef. The high price of butcher-meat, the depressed fortunes of British agriculture, the growing desire of the mass of the people to eat more animal food, the grave importance of the question in relation to public health and the national vigour, combine in inducing us seriously to consider why the flesh of the horse should become an article of daily food. If it be good, it ought to be used; and if it be common in our butcher-markets, this new article of diet must enlarge the sphere of the farmer's labours as the purveyor for the sustenance of the nation. He is no longer a mere grower of vegetable productions: he is becoming more and more a manufacturer of animal food; and there is no good reason why he should only supply us with that furnished by the ox, the sheep, and the pig. Let him satisfy himself as to the agreeable and nutritious qualities of horse-flesh by eating it on the first opportunity. He will thus help to beat down an unhappy prejudice, and by accustoming the people to enlarge the limits of their diet, he will do that which will benefit them, and at the same time extend the field of his operations as a raiser of animal food.

But as few farmers will "eat a horse," either from philanthropy or in hope of filling their purses by inducing their neighbours to become hippophagous, we shall endeavour to stimulate them to perform their office of feeding us on whatever we will eat of that which the land produces, by acquainting them in particular, and the meat-eating public in general, with what is doing on the Continent in this matter of eating horse-flesh.

Horse-flesh is freely eaten at Vienna and in many other Continental cities, but the public sale of it is forbidden in Paris. Though this prohibition be still in force, it is evidently on the point of being removed, the authorities seeing no reason why that which is largely eaten, and with manifest advantage, should not be publicly sold. The last lingering prejudice is about to yield, vanquished by the irresistible logic of facts.

Though the Parisians cannot buy the flesh of the horse in the public markets or in private shops, it is nevertheless an article of considerable consumption. Systematic efforts are being made to extend the use of it; and so far from being sentimental, and listening to the outcry of some who protest against the barbarity of eating an old friend who may have served us faithfully in love or in war, the horse-flesh-eating movement is mainly promoted by the Society for Preventing Cruelty to Animals. At the beginning of last year, this Society appointed a Commission to take all steps in order that the flesh of the horse shall become an article of public consumption.

The Commission was of opinion that the best way of accustoming the poor to the new diet was to offer it gratuitously. It, therefore, opened a subscription for the purpose of buying healthy worn-out horses, and giving them away. The list was soon numerously signed. M. Ducoux, formerly Prefect of Police, and Director of the Cab Company, paid his subscription in kind by presenting a horse to be distributed among the poor at Easter. The Countess of Clérambault, of whose comical conversion to the creed of the hippophagi we have yet to tell, is also to give a horse, so convinced is she now of the importance of affording the poor opportunities of getting rid of prejudice against such valuable food.

It is thus manifest that the zeal, eloquence, and ability of the lamented naturalist, Isidore Geoffroy Saint-Hilaire, are bearing fruit, and that there will soon be the fulfilment of his ardent wishes that the working-classes in France should participate more freely of animal food, not from creatures to be acquired abroad and acclimatised with care and cost, but already existing in France abundantly. France contains more than 3,000,000 horses. It is calculated that these animals are renewed every fifteen years: for at the expiry of this period they are worn out, and must be got rid of.

If we suppose the fifteenth part of 3,000,000 handed over to the butcher, we have 200,000. Suppose 50,000 of these diseased and unfit for food, we still have 150,000 healthy animals, which would furnish 6,000,000 lb. of nutritious food. As that weight is equivalent to the meat furnished by about 100,000 head of cattle, the question is of great economical importance in France, where the working-classes, especially in the country, do not eat butcher-meat more than five or six times a-year. M. Play, in a work devoted to the subject, has demonstrated that a man's daily normal consumption of animal food should be 250 to 300 grammes; and that for the attainment of this rate by the whole French people the quantity of butcher-meat must be three-and-a-half times greater than it is.

"John Bull" is very far from being such a beef-eater as is generally imagined; but poor "Jean Crapaud" is so much of a vegetarian, by compulsion of course, that it is really most wise and benevolent to make a man of him by giving him more of the "strong meat for men," which undeniably is found in the flesh of the horse. The Commission to which we have already referred, acting on the Scotch proverb, "The proof o' the puddin' is the preein' o't," organised last March a grand banquet, at which nothing was served save horse-flesh, in order that amateurs might make themselves acquainted with the merits of this kind of meat. Moreover, a number of Parisian workmen have petitioned the Prefecture of Police to authorise the opening of butcher-shops in which nothing shall be sold but horse-flesh. This demand is reasonable, as a protection against those fraudulent butchers who sell horse-flesh instead of beef, charging, of course, the price of the latter.

An additional proof of the progress of Parisian hippophagy is recorded in recent numbers of the 'Bulletin of the Imperial Society of Acclimatation.' Before a distinguished audience of ladies and gentlemen, members of this most useful society, M. E. Decroix, veterinary surgeon, Guard of Paris, not only read a paper on horse-flesh as an aliment, but supported his arguments by inviting his hearers to judge for themselves by presenting soup, roast, boil, stew, all of horse-flesh.

Our intrepidity in "eating a horse" is as nothing compared with that of this most intrepidly benevolent hippophagist. Engaged in military service in Algeria during the campaign of 1859, he was obliged to sacrifice his horse, which had been seized with paralysis of the hind legs. In his previous campaigns in Algeria, the Crimea, and Italy, he had seen good dead horses constantly allowed to rot, and was no wiser than his comrades who witnessed this shameful waste without thinking of their folly. But the paralysis of his own steed recalling to him the exhortations of the hippophagi, the valorous veterinary surgeon, instead of burying his defunct horse, ate him. And now he declares that to thousands in Algeria and France he has given horse-flesh, to their perfect content, yea, even delight. He groans at the remembrance of the miseries of the English soldier in the Crimean war, while relief was at hand in good horse-flesh. He believes that hippophagy will render great service to armies in the field; and that when horse-flesh is openly sold in shops and markets, soldiers will lose their prejudice against it, and that sailors will not throw wounded horses into the sea, especially when their rations are salt meat producing scurvy. He also justly remarks, that to understand its advantages to an army during a campaign, it should not be compared with the flesh of good animals, but with that of the lean, worn-out, wretched oxen, which follow expeditionary columns.

But our benevolent veterinarian has illustrated the value of horse-flesh to civilians in a way perfectly unique. Something magnanimous was to be expected from a man who ate *his own* horse. We sometimes, in a loving and philosophical—that is, hippophagous—spirit, look on a neighbour's horse, in temporary forgetfulness that we ought not to covet it, though it is said not to be so desirable for food as his ass! But we have compunctious visitings—qualms of the inner man—when we think of our dear old mare appearing at table to be eaten like roast beef. And, even in our most hippophagous mood, we never thought of feeding our household and friendly guests on horse-flesh, regularly supplied *gratis* through the politeness of the proprietor of a slaughter-house. But, apparently with no idea that he had done anything out of the common, M. Decroix informs his refined audience that for two years he had in his own house used only horse-flesh, thanks to an especial authorisation of *M. le Préfet de Police*, and the obliging generosity of M. Macquart, the chief Parisian knacker. And then, by way of a clencher to his philosophical argument, he looks them in the face, complacently remarking—" And I am quite well."

He finds it necessary, however, to admit that the mental state of those eating horse-flesh, and the conditions under which it is presented, must be taken into account. People do not make sufficient allowance for the manner in which it has been kept, and, above all, prepared. A bit of beef may be tender or tough according as it has been cooked immediately after the slaughter of the ox, or kept for several days, and according to the way in which it has been well or ill treated. It is the same with horse-flesh.

As to the mental state of the consumer, in perverting his judgment, we must let M. Decroix " give one proof out of a thousand :"—

"A month ago I went to spend a few days in the Pas-de-Calais, at the country-house of the Countess Clérambault. I had taken with me a stew from a horse distributed to the poor. The day after my arrival at Ligny there was a great dinner. The time for serving up my dish of horse was favourable; and the cook, being let into the secret, presented it with an appropriate sauce. (I knew the guests too well to be apprehensive of stomachic indisposition or mental irritation.) The dish was most successful; everybody ate of it—even a priest from Calais, who said he had no appetite. The countess was helped twice.

"Next day at the family breakfast I announced that, being desirous to make them acquainted with horse-flesh, I had brought a piece, which was to be served up like beef *à la mode*. At this moment a piece of beef was brought in. The countess, not finding the pretended horse very appetising, would not at first touch it; but at my entreaties she took a little bit, which she thought hard, disagreeable, bad, and ordered away with a gesture of disgust.

"'Nevertheless,' said I, 'it seems to me there is not much difference betwixt this horse *à la mode* and the beef *à la mode* which we had at dinner yesterday.'

"'Nonsense,' said she; 'your horse is not so good as the beef.' 'And I,' added a lady from Paris, 'think that the beef at dinner was much better and more tender than your horse.' These positive assertions being enough for my case, I told them of the substitution which had been made, in order to convince them that there was nothing against horse-flesh but silly prejudice. At this revelation all the guests burst out a-laughing, but the substitution appeared so incredible that the cook had to give his testimony in order to remove every doubt.

"'Now,' said the countess, 'that I certainly know the good quality of horse-flesh, I shall have some of it served up at my next dinner-party.' 'And I,' added the lady from Paris, 'when I go back, I shall give it to my best friends to eat.'"

M. Decroix is much too knowing to carry on the war against prejudice merely by the light arms of a practical joke: he attacks it seriously with the heavy artillery of facts and reasonings systematically arranged. As the most recent exposition of the views on public alimentation zealously being propagated in France, we have read M. Decroix's address to the Imperial Society of Acclimatisation, and shall avail ourselves of it freely in bringing before our readers the *pros* and *cons* of the horse-eating controversy.

As to the alimentary qualities of horse-flesh: slaughter and skin a horse for the butcher, there is no notable difference between its flesh and that of the working ox, in respect to colour, smell, and general appearance. When cooked, the resemblance is so great that M. Decroix for more than a year gave it at his own table to his relatives, friends, and acquaintances without their suspecting what it was. Stewed, roasted, *à la mode*, and as soup, it was always pronounced good, sometimes perfect. When boiled, it was generally firmer, and always much leaner, than beef. In this form it is not so agreeable as ordinary beef, but it is more nourishing and salubrious than young oxen prematurely and rapidly fattened, and whose flesh is pale, soft, watery, and too loaded with fat.

As to its salubrity, Baron Larrey, the father of military surgery, often fed the sick and wounded on horse-flesh, and pronounced it perfectly healthy; he even declared that it helped to remove an epidemic of scurvy. M. Baudens fed with it the soldiers of a battery of artillery, and they escaped the serious disorders which were devastating the rest of the army.

In 1856 the Council of Health for the Seine, officially consulted by the Minister of Agriculture, declared that horse-flesh is healthful, and should be publicly sold in a special butcher-market.

In respect to salubrity, reason completely accords with science. We may almost certainly infer what is the alimentary value of the flesh of any species of animal when we know what it feeds on. The general rule is that the flesh of the *herbivora* is salubrious, nourishing, and suited to our functions of digestion. That of the *carnivora* is disagreeable, and impregnated with an odour exciting disgust. There is great difference in that of the *omnivora*, according to the degree in which they have fed on matters animal or vegetable. For instance, that of pigs fattened on flesh furnishes fat—soft, oily, and of indifferent quality; those receiving grain give excellent lard.

Applying these principles to the horse, we conclude that its flesh should be salubrious and agreeable. The researches of several chemists, and of Liebig in particular, prove that its constituent elements are those of beef, with the exception of its having a slight excess of the highly nutritious principle called *kreatine*.

Experience tells us the same. The societies for the protection of animals at Vienna, Berlin, Hamburg, Brussels, Copenhagen, and at other places where horseflesh is regularly sold in special butcheries, all declare that it has led to no inconvenience, and most of them express the wish that the use of it may be extended. So much for what science, reason, and experience declare on behalf of hippophagy. Let us now see what are some of the objections.

If, it is said, horse-flesh were really good, it would have come into use long ago. They who so speak know not what they say, and enable an opponent to retort that it must be good, seeing that, among many nations, it has been eaten for three thousand years; or, to speak more correctly, it has at one time or another been eaten all over the world, as M. Isidore Geoffroy Saint-Hilaire has learnedly and unquestionably demonstrated.

A more serious objection is that the horse is subject to frightful diseases—glanders and farcy, which may be communicated by contagion. But for twenty years the contagiousness of glanders has been doubted by many; and the occurrence of this disease is much less common than formerly. And now so rare is it in Paris, that it is often difficult to procure a case of it for scientific purposes. But if it were as common as it is rare, it is easy to guard against it, just as the public is protected against oxen smitten with carbuncle, a much more formidable disease than glanders.

It is objected that good horse-flesh would be too dear. Parisian horses unfit for work from old age, premature use, and various accidents, are worth about twenty francs; and all the meat eaten by M. Decroix, or given by him to others during five years, was that of animals of the average value of fifteen francs. He insists that it was not merely good, but very good! No doubt a good horse generally costs more than a good ox. Nobody thinks of sending to the butcher horses worth £60, but only those unfit for work, and consequently not worth more than from £2 to £5 at the most. One of the horses given to the poor cost 15 francs, and yielded about 450 lb. net, which is less than a penny a-pound. If the working-classes believe this—and really it cannot reasonably be called in question—it is evidently their own fault if they do not speedily make more acquaintance with the comfort yielded by flesh-pots. We pray them to reflect that every horse buried or given to the dogs is a loss, on the average, of 450 lb. of valuable food. If they fancy that the flesh of an old horse must be black, stringy, tough, and disgusting, experience shows that it is not. Almost all the horses eaten by M. Decroix and his friends, or given away, were aged; and, nevertheless, he adduces the evidence of priests and sisters of charity to show that the meat was sought after with avidity.

Having thus disposed of objections, our author proceeds to speak of the advantages of hippophagy. Here he rides his hobby with consummate intrepidity, scouting the idea that it is cruelty to feed on animals which have served us long and well, and placing first the *benefit to the horses*. It is easy to laugh at the notion of killing horses in order to save them pain, but a little reflection shows that the humane societies which patronise horse-eating are right when maintaining that thus much cruelty is prevented. A labouring ox is thought to be rightly used when, being unfit for work, it is fed for the butcher. Why should it be otherwise with an aged horse? To shelter his old age, feed him generously, and then administer "the happy despatch;" —if the beast were as philosophical as M. Decroix, and could choose the mode of exit from harness and all the horrors of equine senility, what else could he desire? The Vienna Society for the Protection of Animals, taking this view of the matter, and disregarding the sarcasms of thoughtless hard-hearted witlings, purchased, in 1854, 1180 worn-out horses, and had them slaughtered for sale. And so successful have been their efforts, that, in 1860, there were in the Austrian capital seven butcheries, which disposed of 1954 horses, purchased at the average price of about £3.

Some ultra-opponents of hippophagy propose to erect invalid establishments for the reception of old horses. Those who love their fellow-creatures more than the lower animals will not, however, cease their efforts to make horses contribute to the sustenance of man; because, argues M. Decroix, they deem it *lèse-humanité* to lose so much valuable meat which so many human beings need. Upon the whole, then, we think that our benevolent readers, however sentimental, will find it hard to answer M. Fleulard, Secretary of the "Société Protectrice," Brussels, who writes thus:—"Thinking of the fate in reserve for most old horses, which we are pained to see expiring under blows and excessive work, and even starved to death by their barbarous owners, we conclude that it is most misunderstood pity to wish the animals a continuance of such a life rather than take them to the slaughter-house."

There is another consideration partly compassionate and partly prudential. In the last stage of their sad life horses frequently cannot be disposed of from various infirmities, especially blemishes ending in lameness. Nevertheless, their maintenance costs as much as if they were good workers. These poor animals have to undergo painful operations, the application of fire for instance, often resorted to in desperation, and not always able to repair the waste induced by excessive fatigue. This period, so very painful to them, and so costly to their owners, would be shortened by the establishment of horse-meat stalls, offering an advantageous way of disposing of the animals.

The advantages to the proprietors of horses are, according to M. Decroix, equally obvious. An old horse sent to the slaughter-house is at present worth about 20 francs; but when hippophagy becomes general the average price, it is supposed, will be raised to 75 francs; and as France and Algeria possess three millions of horses, and one million of asses and mules (the flesh of which is better than that of the horse), the public fortune will be increased by the sum of £119,046; and the working-classes will have a large amount of horse-flesh at less than twopence a-pound.

It were well that everybody should help to beat down prejudice by, at least occasionally, eating horse-flesh. But as the *name* is not inviting, we propose that the viand should be termed *cheval*, on the same principle that we derive from the French the names respectively employed to designate the flesh of the sheep, the calf, and the ox. It is to be hoped that the altered nomenclature will facilitate the introduction to our tables of a valuable and savoury species of food, and that even the most refined ladies will not look shy at roast *cheval;* and that the fastidious dyspeptic will gladly partake of *cheval* soup, when assured, on the authority of Liebig, that it contains a notable quantity of that remarkable crystallised substance to which he has given the name *kreatine*, and which appears to exercise a singular function in the digestion of food.

But the poor and the working-classes are specially interested in these foreign doings relating to alimentation. In 1862 the poor receiving assistance in twenty districts of Paris were 115,114; to whom were distributed 162,535 francs worth of meat cooked and raw; that is, each of the poor received about 2¼ lb., costing about fourteenpence, in the year. In a capital where butcher-meat is so dear, it is certainly inexplicable that any obstacle should be thrown in the way of obtaining an additional supply. Any increase of agricultural produce, however limited, is hailed as a public boon; but with whimsical inconsistency many of the Parisians still deride the proposal to introduce into the markets the annual addition of about a hundred millions of pounds of excellent animal food. It is most short-sighted policy to under-feed the man who feeds us—that is, the working-man. You may house him better and raise his wages; but to enable him to bear up against toil and sickness you must give him more animal food. "*Pour bien travailler il faut bien manger*," says M. Decroix; which is the French form of the Scottish proverb, "It's the meat that works the wark, and no the lang day." This remark is specially commended to the consideration of agriculturists in Scotland. No class of the community eats so little butcher-meat as that of the ploughman. Fortunately he is not discontented: weekly he sees mountains of beef and mutton transported from the farms where his labour produced that which enabled his master to gratify the carnivorous longings of the community. And yet his teeth do not seem to water! Visions of rump-steaks, rounds of beef, and savoury soups, appear not to impair his appetite for the inevitable *brose*. Well, if he be satisfied, why discompose his mind and derange his digestion, it may be, by cultivating a taste for mutton broth, or *cheval* soup? To those who so speak we reply,—he has an apparatus for masticating and assimilating animal food precisely like your own. And if, as we know is the case, he be no longer fit for the regular work of the farm at the age of little more than fifty years, we think, as a question of social economy, it deserves to be considered whether the sameness of his vegetable diet has not much to do with the comparatively brief duration of his ability to labour.

Science has already answered the inquiry in the affirmative. And as his wages, like those of many others of the sons and daughters of toil, will not permit him to buy much butcher-meat, there seems to us to be no evading the conclusion arrived at by many distinguished foreigners—namely, " Science, reason, and experience declaring that the flesh of the horse is salubrious, agreeable, and highly nutritious, it is most needful that it become a common article of food, with a special view to ameliorating the lot of the poor and the working-classes."

MYCOPHAGY;

OR, SHOULD WE EAT FUNGUSES?*

"DIVIDE and govern" is the political axiom believed to guide the conduct of him who aspires to be what old Homer styles "a king of men." The same axiom must, in another sense, control the conduct of those seeking to make themselves acquainted with the kingdom of nature, and to use the knowledge thus acquired in the advancement of science, and in increasing the happiness and comfort of their fellow-creatures. The field of observation is so immense that we are bewildered by its greatness, and lose ourselves vaguely wandering amid its wonders, until, satisfied with the comparative fruitlessness of a general survey, we set limits to our investigations, and confine them to some particular department attracting us by its importance, or by its being specially within the sphere of our knowledge or our tastes.

It is thus that men acquire precise information, and discipline their faculties for the making of new and

---

* 'Outlines of British Fungology,' by the Rev. M. J. Berkeley, M.A., F.L.S.; with coloured Figures and Dissections of 170 Species, by W. Fitch. 'The Esculent Funguses of England,' by the Rev. Dr Badham; with 20 coloured Plates. 'Illustrations of British Mycology,' by Mrs Hussey; First and Second Series; 140 Plates.

bolder excursions into the regions of the unknown or the little understood. He that intelligently writes a monograph is the possessor of trustworthy information, for which we shall vainly look from the discursive student imitating Solomon, who "gave his heart to seek and search out by wisdom *all* things that are done under the sun." The value assigned to the researches of those known to be devoted to particular studies is manifested by the authority accorded to their opinions, and by the eagerness with which editors of encyclopædias solicit their contributions. Witlings, no doubt, make merry with the labours of the retired student spending his days in elucidating some apparently minute point in geology, botany, or natural history. But the philosopher, recognising the worth of every carefully ascertained fact, fails not to remember gratefully the names of those who have helped to rear securely the temple of science by depositing on its slowly rising walls the materials which they have painfully gathered. Very thoughtless is the wonder sometimes expressed at the enthusiasm exhibited in the exploration of some nook or corner far away from the observation of the unreflecting multitude.

> "Various as beauteous, Nature, is thy face.
> . . . All that grows has grace.
> All are appropriate. Bog and marsh and fen
> Are only poor to undiscerning men.
> Here may the nice and curious eye explore
> How Nature's hand adorns the rushy moor;
> Beauties are these that from the view retire,
> But will repay the attention they require."

These lines of Crabbe we commend to the attention of those—alas! too many—"undiscerning men," who have never asked what is the use of a fungus, and who carry their ignorant contempt so far as to trample under foot, as noxious "toad-stools," numerous species of vegetable productions, not a few of which on examination are found to exhibit the most graceful forms and the most brilliant colours, as well as to supply the most exquisite odours, and abundance of the most palatable and nutritious food. " What geometry," asks Dr Badham, " shall define their ever-varying shapes—who but a Venetian painter do justice to their colours? As to shapes, some are simple threads, like the byssus, and never get beyond this; some shoot out into branches, like sea-weed; some puff themselves out into puff-balls; some thrust their heads into mitres; these assume the shape of a cup, those of a wine-funnel; these are stilted on a high leg, and those have not a leg to stand on; some are shell-shaped, many bell-shaped; and some hang upon their stalks like a lawyer's wig; some assume the form of the horse's foot, others of a goat's beard; the *Phallus impudicus* is the very thing he calls himself. As to their colours, we find in one genus only species which correspond to every hue!" As to odours, while some smell like cinnamon, some like ratafia, and some "like the bloom of May," Dr Badham, enthusiastic though he be, cannot discredit his nose, which is unmistakably and instinctively turned away from fungous odours yielding " an insupportable stench," " an intolerable fœtor," " the savour of a stale poultice," " a smell of tallow," " the smell of putrid meat."

When our readers bear in mind that very many fungi are violently poisonous, so that Dr Badham suffered severely from merely tasting one of the spores of the milky Agarics which he had collected, they may think that there is ample justification of the popular aversion to the whole fungus tribe. We presume to differ from the popular verdict. As, in the interest of public alimentation, we lately besought favour for hippophagy, and declared that the dining on soup and stew made from horse-flesh caused us no digestive remorse, we have now entered on a course of practical mycology by eating puff-balls—*Scotticè*, " deil's sneeshin"—of which our palate much approves, and against which our stomach testified no displeasure. To eat a horse is, no doubt, much less enterprising than to feast upon funguses, many of which are virulently poisonous, and some of which vary in their qualities, so that what is safe here now may be the reverse at another time elsewhere. Although, then, as the French newspapers express it when speaking of doubtful political rumours, we must commend funguses "with a certain reserve," we nevertheless hope to interest our readers by telling of their good qualities as well as their bad, and explaining their not unimportant functions in the wise economy of nature.

Being passably honest, we hope, and not wholly without prudence, as we verily believe, we do not wish to hide from ourselves or our amiable readers that the mycophagist may pay dearly for gratifying his palate, even though his hastily-summoned medical attendant may not pronounce him moribund; and therefore we have not only alluded to Dr Badham's sufferings, but now also warn those seeking acquaintance with funguses, that the mere tasting of some of them experimentally will produce contraction of the jaws, sickness, pain and heat in the stomach, as well as slight delirium. Bad may go to worse, and the unhappy inquirer may be afflicted with giddiness, debility, loss of sight and recollection, burning thirst, vomiting, fainting, and violent gripes—nay more, a man who has been poisoned by eating funguses may not know his perilous condition until suddenly made aware of it, many hours after partaking of his last fatal meal, when it is too late to adopt measures to eliminate the poison. We thus chronicle the perils of mycophagy, because Dr Badham's praises thereof are so eloquent and so witty, and the beautiful delineations of his friend Mrs Hussey are so inviting, that we fear some wanderer amid our summer fields or autumnal glades may be too rash in tasting of the lauded fungi. In fact, in a particular instance, when comparing Badham's figure of a certain species with that figured in the admirably illustrated work of

Berkeley, the discrepancy was so evident that we thought it prudent to give our stomach the benefit of the doubt. Had we been more enterprising, it might have been a case of death in the pot, chronicled in the newspapers as a melancholy instance of over-confidence in a Hussey.

It must then be acknowledged that we have here an instance of the pursuit of knowledge under difficulties. With Berkeley or Badham as our "guide, philosopher, and friend," we might gain such speedy knowledge of the good and evil that are in funguses, as not to fear feasting on those recommended. But with only printed directions, however explicit, and pictorial illustrations, however faithful, we think our prudence praiseworthy in hitherto confining our gastronomical acquaintance with fungi to three varieties of mushroom and two of puffballs.

If some friend, practically a mycophagist, have the kindness to put into our hands the varieties of fungi which he knows to be wholesome, or if he eat of them before us in order to remove our scruples, it may be all right to "dine with what appetite we may" on what generally excites fear and loathing. It is really not safe to follow one's nose, or believe one's eyes, as the saying is. All noses are not equally furnished with olfactories, it would appear, and not a few are in some strange fashion abnormal. We know a lady who thinks candle-snuff a delicious odour! And as to following optical appearances in deciding on the edibility of a fungus, who does not know that many of us are "short-sighted," and that a considerable percentage of human eyes are afflicted with colour-blindness? We may be told that Patagonian savages and Russian boors thrive on funguses of which we never venture to taste. Very true, we admit; but then the savage does not wear spectacles, and has "senses exercised by reason of use" in a way unknown to the civilised savant, whose artificial life does not require such absolute dependence on the reports of his perceptive organs. In a matter of mere seeing or smelling we would follow the Patagonian savage rather than the Pope, or the President of the Royal Society.

In short, like other doubtful characters, let not funguses be trusted and admitted to our tables until our doubts be removed. But caution need not be so excessive as to shut us up endlessly in insular prejudice. In the 'Gardeners' Chronicle,' edited by Dr Lindley, we are astonished to read such a sentence as this:—"We admit that we throw away many sorts that are excellent, and that we thus deprive our palate of much gratification, and perhaps that those in common use are inferior to those that are rejected. But we err on the safe side; the mushroom, chantarelle, champigny, morel, and truffle, are enough for luxury; and we trust that the peasantry of Great Britain will never be brought to the condition of Italian lazzaroni."

This looks very like "bunkum" to flatter John Bull. The question is not one of "luxury" for the favoured few who can afford to pay dearly for morels and truffles, but of food for the many. Ought we to throw away many sorts of funguses on which the peasantry of other nations feed and fatten? It is a mere popular myth that "John Bull" chiefly lives on animal food. Poor fellow! he is, as we have formerly shown, very much of a vegetarian, to which "his poverty and not his will consents." And when by-and-by we enlighten him as to the chemical constituents of the edible funguses of Britain, we are sanguine that no 'Gardeners'-Chronicle'-inspired contempt for the lazzaroni will keep him from coveting an article of food which in savour is not merely like his national beefsteak, but, to a greater extent than anything vegetable, is actually composed of the same elementary substances. If an Italian or a Russian peasant commit wonderfully few mistakes as to the quality of fungi, can nothing be done to educate our people so that it shall be safe for them to make these productions common articles of diet? We think it more patriotic to try to solve that question than to help to keep up the silly notion that the British peasantry would sink in the social scale by so far resembling foreigners as to feed upon funguses.

We shall, therefore, with the help of the works already referred to, give the outlines of Fungology, and make special mention of those British fungi which are valuable as esculents.

Fungology, as a scientific term, is objectionable, because combining a Latin with a Greek word. The spurious term is, however, very generally received. The fastidious classical purist may please himself by substituting mycology, which is at once etymologically correct and sufficiently comprehensive. The word fungus may, however, in any case be retained in common parlance, only, as Berkeley beseeches, we must not display our lack of learning by speaking of 'a fungi,' as in Phillip's 'Prize Essay on the Potato Murrain,' for this is "grating to the ear, and utterly intolerable. If fungus be considered an English word, as it is by some of our older authors, the plural will be funguses; but there is, then, something unpleasing in the sound, and the term fungi is certainly to be preferred."

Some melancholy etymologists, upon whom Badham thinks good mushrooms really thrown away, derive fungus from *funus*. The Dutch, on the other hand, from supposing, no doubt, that the devil eats the best of everything, compliment funguses with the title "Duvyel's broot."

As to the origin of the term toad-stool, Badham asks, "Have not the words *tode*, and the stool called after him, some etymological, as they have undoubtedly a fanciful, connection with the word *todt* (German), death?" Very likely, we think; but then our learned author does not perceive that his query favours the theory of the melancholy etymologists who associate fungi with funeral. "The origin of the word toad-stool," he goes on to observe, " which makes them seats or thrones for toads, does not quite satisfy me, I confess, though there be doughty authorities for it in Johnson's Dictionary, and in Spencer's 'Faery Queen'!—

> ' The grisly todestool grown there 'mought I see,
> And loathed paddocks lording on the same;'

and though an anonymous Italian authority declares that in Germany they have actually been seen sitting on their stools, still even in Germany it must be admitted that they do not use them as frequently as we might expect, had they been created for this end."*

As to the plants comprehended under the general term Fungus, it is not easy to give a strict definition comprehending every individual genus and species of the whole group—a difficulty the origin of which is apparent when it is remembered that it is often extremely difficult to distinguish a plant from an animal, and that the fungus family is not only very prolific, but almost cosmopolitan in its diffusion. Merely catalogued and described, there are sufficient to fill an octavo volume of nearly four hundred pages of close print of British species alone. Declining, therefore, anything like such strict definition, we shall indicate a few of the

* *Apropos* of toads, here is something that astonished us the other day. Taking home a little toad for the purpose of trying whether, when come to years, he could manage to live a few years without food, according to popular belief, we found him covered with loose scales, apparently of silver. Unfortunately our visions of fortune from a silver-bearing toad were dissipated by the speedy death of toadie while we were benevolently considering how to feed him. The scales are in our possession, however—a singular memorial of the defunct. Can this be the foundation of Shakespeare's

> " Uses of adversity,
> Which, like the toad, ugly and venomous,
> Wears yet a precious jewel on his head"?

Last summer, in the museum at Elgin, we found a toad, said to have been exhumed from solid rock, covered with similar metallic scales. What are they?

curious plants comprised in the study of mycology, with the view of affording some general notion of what it comprehends.

Taking the common mushroom as our point of departure, we have the type of an enormous group characterised by a bonnet-shaped receptacle (*pileus*) supported by a stem, and furnished beneath with a number of gill-like plates (*lamellæ*), which, when placed on paper, emit a vast quantity of dust-like bodies, to which, though reproductive, the name of spores has been given, to distinguish them from seeds which contain an embryo, while these consist of a two-coated cell without the slightest trace of an embryo. These spores are of different colours in different species, very frequently pure white, but presenting also pink, various tints of brown, from yellowish and rufous to dark-bistre, purple-black, and finally black. As these colours are accompanied by peculiar differences of habit, they afford a ready test for grouping the species, some of which have the brightest rainbow hues, combined with the most elegant and delicate forms; while others are coarse, dull in colour, and unsightly, few of them being persistent, and many, when decayed, pass into a loathsome mass, in which riot those insects which are nature's scavengers.

"The gill-bearing fungi are generally of a soft substance, but they are not all so. According to the density with which the cells or threads of which they are composed are packed, they present various degrees of hardness, till they assume even a corky substance, and are more or less persistent. The common fairy ring champignon (*Marasmius oreades*) is a familiar example of the first departure from the common mushroom type, and, in consequence of its less watery character, is easily preserved for culinary purposes. The dædalea of the birch (*Lenzites betulina*) gives a good example of the further hardening of the gills, while in that of the oak (*Dædalea quercina*) the substance is as firm as cork, or in parts hard as wood."—(*Berkeley*.)

In a very important group of fungi the spores are the essential character, as the gills are in those just described. In their multitudinous species the polypori exhibit every gradation, from great succulence to the hardness of wood. The scaly polyporus, so common on ash —the coriaceous *Polyporus versicolor*, with its velvety pileus and many-coloured zones, so common on stumps and felled wood—and the hard hoof-shaped polyporus abounding in plum orchards—are familiar examples.

The existence of prickles, or spine-like processes, on the under side of the pileus, is the characteristic mark of a third subdivision, called *Hydnei*, after the typical genus hydnum (from the Greek word for truffle). The pretty *Hydnum auriscalpium*, common upon fir cones, attracts attention from the elegance of its form and colouring; and the esculent *H. repandum* is a common inhabitant of our woods.

The characteristic of a fourth group is the absence of projections or depressions on the hymenium or fructifying surface. The species is very common and widely diffused, and is termed *Auricularini*, from some of the most characteristic being ear-shaped. Familiar specimens are to be seen upon the oak-trunk, felled and peeled, in the form of bright yellowish fungi; while a felled poplar, with the bark on, is adorned with a beautiful and somewhat similar lilac fungus.

Hitherto all the species described have something in the shape of a pileus; but in the next group—that of *Clavati*—so named from their club-like form—the pileus disappears, and is replaced by a club-shaped receptacle covered with the fructifying surface. If the stem is branched, we may have every variety of tree-like form. The yellow *Clavaria fastigiata* of our meadows, and the white candle-like bundles, *Clavaria vermiculata*, so common on our lawns in autumn, are well-known examples. Here again we have the most beautiful colouring, though several of the finest European species have not yet been noticed in our woods.

The remaining group of allied fungi is distinguished by the predominance of the gelatinous element, and is that of *Tremellini*, from their soft, flaccid character. " Rotten sticks in our hedges or woods often present bright, tremulous, gelatinous masses of bright orange, purple, or dark brown, which at once attract our notice, while the trunks of the elder and some other trees afford ear-shaped, flaccid masses, which almost escape notice when dry, but with the first shower are exposed to the most careless observer. Sometimes, again, on an old stump, or at the base of a living oak, enormous masses are found, resembling the convolute intestines of some animal, but distinguished by their rich ferruginous or yellowish tints." A familiar specimen of the group is known as *Jew's-ear*. "These six groups form subdivisions of one great association of fungi, characterised by their hymenium being more or less exposed, and at the same time bearing naked spores attached to the lips of certain cells called sporophores. The general name of the division is *Hymenomycetes*, the hymenium being the prominent character."—(*Berkeley*.)

The second great division of fungi is characterised by the concealment of the fructifying surface till the containing sac is burst for the dispersion of the spores. It is termed *Gasteromycetes*. The puff-balls are the best-known example. We merely indicate its subdivisions, with popular examples of each:—

  a. *Hypogaei*—Red Truffle of Bath.
  b. *Phalloidei*—Common Stinkhorn.
  c. *Trichogastres*—Puff-balls.
  d. *Myxogastres*—Dust-fungus of tan-pits.
  e. *Nidulariei*—Bird's-nest Peziza.

We next come to a large division of fungi, generally devoid of beauty, and so minute as to appear to the naked eye mere black specks upon leaves, twigs, &c. Its general name is *Coniomycetes*, from the dust-like nature of the spores. Of four of its divisions we cannot indicate any popular representatives; but with the remaining two the British farmer is unhappily familiar, under the names of rust, smut, and mildew. The *Pucciniaei*, comprising the wheat mildew—the mehl-thau, or meal-dew, of the Germans—are distinguished by their articulate spores; while the *Cæomacei*, containing the bunt and rust, are truly parasitic, dust-like fungi.

The next great division of fungi, consisting of those moulds which bear naked fruit, is termed *Hyphomycetes*, from their filamentous character. Its five subdivisions find popular examples in insect club-mould; scarlet tubercularia; carbonised moulds; blue moulds—yeast and vinegar fungus; and yellow boletus mould.

This terminates the first series of fungi, consisting of four divisions, in which the fructifying bodies are naked and exposed. In the term there are included other plants which differ greatly in structure, but many of which are readily recognised as true fungi, while others are very minute and obscure. These productions, instead of naked spores, have fructifying bodies (*sporidia*) enclosed in sacs (*asci* or *sporangia*). Hence fungi of the second order are characterised as sporidiferous.

The first group comprehends six divisions, whose learned names we shall not inflict on our readers, who doubtless will be satisfied to be told in plain English that these divisions are respectively represented by morel, truffle, maple-mould, candle-snuff fungus, hop-blight, and hoof-fungus. The second group, comprising two divisions, is popularly represented by felt-moulds and bread-mould.

As to the nature of the productions which have been thus classified, they are undoubtedly true vegetables in the main principles of their growth and structure, and divisible into species as definite as in other acknowledged parts of the vegetable kingdom. Berkeley therefore thinks it superfluous to consider the notions of those who regard them as the creatures of chance, or of

ORIGIN OF FUNGI.

a happy concurrence of circumstances favourable to their growth from inorganic elements. "The notion of equivocal or spontaneous generation is now all but exploded amongst scientific men. The most careful experiments show that, without pre-existent germs, no organised beings are ever produced from such solutions as contain matter fit to nourish minute animals or vegetables, though, where proper precautions have not been taken to exclude the possibility of their access, they exist in myriads." Badham, if not so scientific, is certainly more amusing than Berkeley, and thus expresses his hearty contempt for those who fancy that his beloved fungi may originate in "spontaneous" or "equivocal generation:"—

"We might as well talk of the pendulum of a clock generating the time and space in which it vibrated, as of dead matter spontaneously quickening and actuating those new movements of which some of its particles have become the seat; for how, in the name of common sense, can that which we assume to be dead—*i.e.*, emphatically and totally without life—convey such purely vital phenomena as those of intus-susception and growth, which, by the very supposition, are no longer within itself? Life, on such a hypothesis as this, ceases to be the opposite and antagonistic principle to death, of which it then becomes but a different mode and a new phasis. At this rate, addled eggs, abandoned by the vital principle, might take to hatching themselves!"

As to the habitats of fungi, Berkeley soberly declares that it is difficult to point out any substance or situation, where conditions exist capable of supporting vegetation, in which fungi, in one or other of their forms, may not be developed. The rhetorical and facetious Badham exclaims—

"Where are they not to be found? Do they not abound, like Pharaoh's plagues, everywhere? Is not their name legion, and their province ubiquity? To enumerate but a few, and these of the microscopic kinds

—the *Mucor mucedo*, that spawns upon undried preserves; the *Ascophora mucedo*, that makes our bread mouldy; the *Uredo segetum*, that burns Ceres out of her own corn-fields; the *Uredo rubigo*, that is still more destructive; and the *Puccinea graminis*, whose voracity sets corn-laws and farmers at defiance, are all funguses.

"When our beer becomes mothery, the mother of that mischief is a fungus; if pickles acquire a bad taste, if ketchup turns ropy and putrifies, funguses have a finger in it all!

"Some love the neighbourhood of burned stubble and charred wood; some visit the sculptor in his studio, growing up amidst the heaps of moistened marble-dust that has caked and consolidated under his saw. The *Racodium* of the low cellar (*vide* the London Docks, *passim*, where he pays his unwelcome visits, and is even more unwelcome than the exciseman) festoons its ceiling, shags its walls, and wraps its thick coat round our wine-casks. The close cavities of nuts afford concealment to some species; others, like leeches, stick to the bulbs of plants, and suck them dry; these pick timber to pieces, as men pick oakum. They also attach themselves to animal structures and destroy animal life: the *Oxygena equina* has a particular fancy for the hoofs of horses and for the horns of cattle, sticking to these alone. The disease called 'Muscadine,' which destroys so many silk-worms, is also a fungus, which in a very short time fills the worm with filaments very unlike those which it is in the habit of secreting. The vegetating wasp is another mysterious blending of vegetable with insect life. Lastly, and to take breath, funguses visit the wards of our hospitals, and grow out of the products of surgical disease."

In addition to all this, we must add certain marvels not noticed by Badham. One of the most curious properties of certain fungi is their capability of growth in substances which are in general destructive to vegetables. Tannin is one of these, and yet a tan-pit is the habitat of a certain fungus; more than one species is developed on extracted opium, and the factories in India have suffered greatly from their presence. A few years since, a little mould developed in the solution of copper used for electrotyping in the department of the Coast Survey of Washington, proved an intolerable nuisance.

The farmer groans over the mischief occasioned by the potato-mould; the *gourmand* anathematises the havoc which fungi make among the choicest cooked provisions; and the rarest wines prized by the *gourmet* acquire a taste and odour rendering them undrinkable, owing to the presence of a fungus which first attacks the corks. The dry-rot is not merely the terror of naval architects, but occasionally obstructs a railway tunnel! In Dr Carpenter's 'Elements of Physiology,' it is related that in the vicinity of Basingstoke a paving-stone, weighing 83 lb., was raised out of its bed an inch and a half by a mass of toad-stools, and that nearly the whole pavement of the town was displaced from the same cause. And as fungi sometimes occur on glass, and even on smooth metallic substances, the most recent specimens of wonderful architecture—the Crystal Palace and the "Black Prince"—may yet be doomed to furnish demonstration that neither glass houses nor iron ships are safe from the destructive effects of fungi. When we add that fungi find their way into eggs, and that in an incredibly short time they often appear in the very heart of a loaf, our readers will be prepared for the transition which we now make to the diseases caused by fungi.

Spores of fungi have been detected, apparently uninjured, in the dust of the trade-winds, in flakes of snow collected from the air, on the mucous surfaces of the internal organs of animals, and in the dejections of cholera. M. Charles Robins's 'Natural History of the Vegetable Parasites Growing on Man and other Animals' is a large treatise on the diseases occasioned by fungi; and there are many scattered memoirs on the same subject. Berkeley thinks there is no reason to believe that they induce epidemic diseases—such as cholera or influenza — according to an opinion once somewhat prevalent; but a curious production called Sarcina, from its resemblance to minute woolpacks, is a rather constant attendant on cancerous affections of the stomach; and the influence of fungi in producing certain cutaneous disorders is now undoubted. A few spores rubbed into the skin, or inserted in it, soon produce the disease termed *Porrigo lupinosa;* and recent experiments tend to show that this direct influence is greater than has been generally suspected. Dr Lowe has induced skin diseases by inoculation with the granules of yeast, and is disposed to attribute a great deal more to the agency of fungi than has been hitherto allowed; and the knowledge of their influence, whether externally or internally, has led to it being counteracted by the administration of salts fatal to fungal growth.

The diseases produced by fungi on the staple vegetable food of man are chiefly referable to the lower orders of the tribe. "The spawn, however, of higher species is often fatal to trees and herbaceous plants, by inducing decay among the roots. It has long been known that trees would not in general flourish where others had grown before, and this was attributed to exhaustion of the soil; it is now, however, ascertained that the evil arises from spawn attached to old decaying roots. A most striking instance occurred lately in the Gardens at Kew. Two deodaras were planted before the Director's house, within a few yards of each other, under apparently similar circumstances. After a time, one of these became unhealthy, and it was suggested that the roots should be examined. A scrutiny took place, when it was found that an old cherry-tree formerly stood on the same spot, that its roots were covered with spawn, and that this had extended to the roots of the deodara. The remains of the old cherry-tree were grubbed up, and the diseased portions of the deodara removed, and now it bids fair to thrive without any further check. Herbaceous plants, such as strawberries, suffer from the same cause; and it is now certain that wherever fragments of wood or sticks exist in manure, whether in the garden or field, there is considerable danger. The formidable larch-rot, which converts the trunks of larches into hollow pipes, is often attributable to this cause."—(*Berkeley*.)

Of fungi attacking timber employed in the construction of houses and ships, the most formidable, perhaps, is the dry-rot, which converts the wood into a dry powdery mass, though both it and the wood are often sprinkled with large drops of moisture. In domestic buildings, where little care is exercised in the selection of timber, the wood is often deeply impregnated with spawn before it is used; and we lately visited the house of a friend which had been abandoned after years of discomfort from the presence of this pestiferous visitant. It is some consolation to know that its attacks may be prevented by impregnating the pores of the wood with gas-tar, sulphate of copper, or some other poisonous metallic salt; and that, when established, its progress may be greatly modified by repeated washing with a saturated solution of corrosive sublimate.

Many fungi prey on the tissue of living leaves. The hop mould, the rose mildew, the vine mildew, and other allied fungi, exhaust the plant and impede its circulation, partly by feeding on its juices, and partly by clogging up its pores. Most of these yield to sublimated sulphur, if timely and judiciously applied. The modes in which these fungi are propagated are so many, however, that they spread with frightful rapidity. The cultivation of the vine in Madeira has almost ceased from this cause, and is everywhere precarious. "It is curious that this fungus has never been found on the American vines, or their numerous varieties, when cultivated in Europe. The Isabella, for instance, a grape of American origin, has always been free from mildew. But though the varieties which are strictly American do not suffer, European kinds imported into the United States frequently suffer."

Berkeley is one of those who attribute the potato murrain to a mould whose spawn attacks the tissues in every direction, being present in the tubers and stems as well as in the leaves. Its spawn never being superficial, the sulphur remedy is unfortunately inapplicable; and for the same reason all external applications, such as lime, are equally worthless. Early planting and the destruction of the haulm immediately after the appearance of the fungus, give the best prospect of success.

It is unfortunate that the word mildew (*mehl-thau*, meal-dew) should be applied to any save the white leaf-moulds, seeing that its application to a particular disease of wheat is constantly inducing error. The diseases produced by fungi with loose, dust-like fruit, and popularly known as smut, bunt, mildew, rust, are many and most injurious; smut and bunt attacking the tissues of the seeds, their floral envelopes, or the receptacle in which the flowers grow, or sometimes the leaves and stems, converting them into a mass of loathsome, and sometimes fetid, dust; whereas mildew and rust attack the leaves and stalks more especially, forming little rusty streaks or spots, and exhausting the plant by the growth of its spores and spawns at its expense. Bunt may be easily extirpated, as the spores, being lighter than water, may be removed by simple washing, and are destroyed by various chemical compositions. The most efficacious remedy for bunt in wheat is perhaps that used in France—viz., steeping the grain in a strong solution of Glauber's salt (sulphate of soda), and then dusting it with quicklime, which coats the seeds with sulphate of lime or gypsum, and sets free encaustic soda for the destruction of the spores.

Under the name of ergot, another fungus produces disease in the grains of rye, barley, wheat, and many field-grasses, converting them into a firm mass without any appearance of meal. Though useful from its medical properties, the prevalence of ergot causes cattle and sheep to slip their young, and, in addition to the fatal gangrene which it occasions in man when entering largely into bread, it is the probable cause of many diseases among cattle when they eat it in seasons during which it abounds.

We have drawn up such a heavy bill of indictment against fungi that we doubt not our readers are curious to know how the charge of irremediable mischief is to be parried or contradicted. We shall, therefore, now proceed to the uses of fungi. By their fermentative and putrefactive powers they perform important functions in the economy of nature. Decomposing even the hardest vegetable substances, they provide a rich supply of vegetable mould for coming generations, besides destroying those structures which, having served their purposes, need to be removed. Several years ago a row of large ash-trees near our habitat fell under the woodman's axe: we have been interested watching the progress by which the unsightly stumps have been removed. They are yearly covered with large specimens of *Polyporus squamosus*, several of them measuring two feet across, while yearly decay makes growing inroads into the formerly solid wood, which now readily yields to a push from a walking-stick, or is converted into a rotten dust which the wind scatters across the neighbouring fields.

An important use is made of a particular condition of certain species of mould in the preparation of fermented liquors, under the form of yeast, which consists of bodies more or less oval, continually giving off joints so as to produce short, branched, necklace-like threads. These joints soon fall off, and rapidly give rise to a new generation, which is successively propagated till the substance is produced under the name of yeast. The globules of which this is composed retain their vegetative powers for months, and can be preserved in a dry state, in which form they are imported as "German yeast;" a compound of capricious deportment, inasmuch as a sudden blow is said to destroy its powers of germination. A still more singular form of mould is the vinegar plant, which, under proper conditions of temperature, has an extraordinary effect in promoting acetic fermentation.

Polyporus squamosus makes an excellent razor-strop, a fact to be noted by Mr Mechi for behoof of *gens barbata*. The German tinder, or amadou of commerce, so familiar to cigar-smokers, is made from the pileus of *Polyporus fomentarius*, beaten out and steeped in a solution of saltpetre. *Polyporus igniarius*, when pounded, is used as snuff by the natives of the northern region of Asia. When burnt, several species of puff-ball have anæsthetic properties similar to those of chloroform, so that operations have been successfully performed under its influence; and it is used in the stupefaction of bees when the object is to remove the honey without destroying them. *Agaricus muscarius* is employed, both in the fresh and dried state, in the production of intoxication, and more profitably as a decoction to destroy bugs and flies. *Cardiceps sinensis*, as stuffing to roast-duck, is said to have wonderfully strengthening qualities, which, Berkeley shrewdly opines, probably reside in the savoury vehicle.

From the bright green produced in fairy-rings by the decayed fungi of last year's growth, it has been conjectured that fungi might, when abundant, form a valuable manure. If collected for this purpose, they should be filled up with alternate layers of sand or light soil to absorb their redundant moisture.

But the great value of fungi is their astonishing resemblance to animal food, and their consequent usefulness as an article of human diet. Of all vegetable productions they are the most azotised—that is, animalised—in their structure. Chemistry demonstrates that they yield the several component elements of which animal structures are made up; and many of them, in addition to sugar, gum, resin, a peculiar acid called fungic acid, and a variety of salts, furnish considerable quantities of *albumen, adipocere,* and *osmazome,* the principle which communicates its peculiar flavour to meat gravy.

Dr Marcet has also proved that, like animals, they absorb oxygen largely, and disengage in return from their surface a large quantity of carbonic acid, and that in lieu of the latter many give out hydrogen, and some azotic gas. Badham is eloquently vehement in denouncing our folly in rejecting an article of food the constituents of which are so nutritious and savoury; while even the calmer Berkeley regrets that in this country we should be surrounded by so large a quantity of wholesome and pleasant food, of which we cannot avail ourselves from mere ignorance.

"The common mushroom, the truffle, and morel, are valuable articles of commerce; but more especially the first, whether in a fresh state or in the form of ketchup. The extent to which this latter article is prepared is quite astonishing. A single ketchup merchant has, at the moment at which I write, in consequence of the enormous crop of mushrooms during the present season, no less than 800 gallons on hand, and that collected within a radius of some three or four miles. The price of mushrooms for ketchup, in country districts, varies very greatly in different years. In the district in which I write, it has not in the present year reached a penny per pound, while in some years as much as fivepence is readily given. In years of scarcity, almost any species that will yield a dark juice is without scruple mixed with the common mushroom, and, it should seem, without any bad consequence, except the deterioration of the ketchup. The best kind is undoubtedly made from the common mushroom (*Agaricus campestris*), and especially from that variety which changes to a bright red when bruised. That from the champignon is excellent, but so strong that it requires to be used with caution."

Even when confessing that he is not so enthusiastic a mycophagist as his friends Mrs Hussey and Dr Badham, Berkeley subscribes to their views as to the advantage derivable from the use of many species of fungi. In fact, he attributes the injurious effects frequently exhibited, not so much to their poisonous qualities as to the gluttonous appetites of those devouring them. A man, after a day's fast, eats a pound or two of mushrooms badly cooked, and frequently without a proper quantity of bread to secure their mastication, and is then surprised that he has a frightful fit of indigestion. Those desirous to practise salutary mycophagy must partake of the tempting vegetable beef-steaks in moderation, and eat plenty of bread along with well-cooked and perfectly sound fungi. A half-rotten mushroom spoiled in the cooking, moreover, is not "a dainty dish to set before the king;" and we may be sure that it was not such an unsavoury morsel that, by Agrippina's directions and Locusta's cookery, had the honour to administer "the happy despatch" to the Emperor Claudius.

Knowing that perhaps no country is richer than ours in excellent fungi, more than thirty species abounding in our woods, we cannot but regret the popular unacquaintance with what Roques styles the manna of the poor. In France, Germany, and Italy, funguses not only constitute for weeks together the sole diet of thousands; but the residue, either fresh, dried, or variously preserved in oil, vinegar, or brine, is sold by the poor, and forms a valuable source of income to those who have nothing else to bring to the market. So thoroughly ignorant of all this are we, that of our Scottish readers we question if a dozen ever tasted a fungus preserved in any of those various ways by which its dangerous properties are greatly diminished. There is some excuse for this. There is no British Inspector of

Funguses to look after the safety of the London lieges; but the "Ispettore dei Funghi" at Rome is an important functionary, as our readers will perceive from one of his rules for the fungus market: "The stale funguses of the preceding day, as well as those that are mouldy, bruised, filled with maggots, or dangerous, together with any specimen of the common mushroom detected in the baskets, shall be thrown into the Tiber."

We think a few species of mushrooms the only safe fungi, while at Rome their character is so suspicious that they are under a Papal ban. We have a few things to learn yet, it would appear, not from the Pope alone, but also from the Emperor of China, whose care of "the Flowery Land" is evinced in the yearly gratuitous distribution of a six-volume treatise, entitled 'The Anti-Famine Herbal.' It contains descriptions and representations of 515 different plants, whose leaves, rinds, stalks, or roots are fitted to furnish food for the people. Not knowing the language of the Celestial empire, we cannot depone as to the funguses eaten by the omnivorous Chinese; that their number is legion we doubt not.

On the authority of Badham we state that the annual revenue of Rome is benefited by the sale of fungi to the amount of at least four thousand pounds sterling. What, then, must be the receipts from funguses throughout all the marketplaces of all the Italian states!

We shall indicate a few British species with which it will be for their advantage that our readers make themselves acquainted. *Agaricus prunulus*, growing in rings about the same time in spring, not in autumn, like most fungi, is generally destroyed by the British farmer, as injurious to his grass crops. In the Roman market it easily fetches 15d. a pound, and is sent in little baskets as presents to patrons, fees to medical men, and bribes to lawyers. Tolerable proof this of its edible value! Badham places it first in his series of plates, as the most savoury fungus with which he is acquainted.

Agaricus procerus, growing abundantly in autumn, and occasionally through the summer. Were its excellent qualities better known, they could not fail to introduce it to a prominent place in the British *cuisine*. Ketchup made from it is much finer than that from the common mushroom.

Boletus edulis, though much neglected in this country, Berkeley pronounces a most valuable article of food. "*Nihil quod tetigit non ornavit*," exclaims the enthusiastic Badham. It imparts a relish alike to the homely hash and the dainty ragout. In Hungary it is made into soup. Here is the recipe: Having dried some boletuses in an oven, soak them in tepid water, thickening with toasted bread till the whole be of the consistence of a *purée;* then rub through a sieve, throw in some stewed boletuses, boil together, and serve with the usual condiments. We have found this delicious boletus in abundance close to our habitat in Forfarshire, and feast upon it freely.

We must not linger over Badham's savoury recipes for mushrooms. That *á la* Marquis Cussi is not just the thing for the multitude, seeing that it involves simmering in a couple of glasses of sauterne. But a homely mode of cooking "buttons" is to cut them with bits of bacon the size of dice, and then to boil them in a dumpling; to master which with comfort would, we fancy, require the digestive vigour of a ploughman's stomach.

Agaricus exquisitus, called also the horse-mushroom, from the enormous size to which it sometimes attains (5 lb. 6 oz., for instance), is generally shunned by the English epicure; and with reason, it would appear, for both Berkeley and Badham hold its tempting name a misnomer.

Agaricus deliciosus is in more favour with Badham, who pronounces it one of the best agarics with which he is acquainted. As it grows somewhat abundantly under old Scotch firs from September to the beginning of November, our readers may test his verdict—"firm, juicy, sapid, and nutritious."

Agaricus oreades, the champignon, or fairy-ring mushroom, is much to be commended—for two reasons, the facility with which it is dried, and its very extensive dissemination. Moreover, a three days' exposure to the air dries it; and in this state it may be kept for years without losing any of its aroma or goodness. In the French *à la mode* beef-shops in London it is used with the view of heightening the flavour of that dish; and it is famous for the flavour it imparts to rich soups and gravies.

Cantharellus avarius.—Battara, in his 'Fungorum Historia,' confesses that some think it pernicious; but he depones that he has eaten it greedily, and insists that, made into soup, it will revive the dead! "The very existence of such a fungus at home is confined to the freemasons, who keep the secret! Having collected a quantity at Tunbridge Wells, and given them to the cook at the Calverley Hotel to dress, I learned from the waiter that they were not novelties to him; that, in fact, he had been in the habit of dressing them for years, on state occasions, at the Freemasons' Tavern. They were generally, he said, fetched from the neighbourhood of Chelmsford, and were always well paid for. Of the *cantharellus*, this summer, the supplies were immense."—(*Badham.*) *Morchella esculenta*, or the morel, is the expensive luxury which the rich procure at great cost from the Italian warehouses, and the poor are fain to do without. And yet it is not infrequent in our orchards and woods during the beginning of summer. Roques reports favourably of some specimens sent to him by the Duke of Athole; and others from different parts of the country occasionally find their way to Covent Garden.

Fistulina hepatica, "the poor man's fungus," as Schöeffer calls it, deserves, indeed, the epithet, if we look to its abundance and its enormous size, Badham having picked a specimen weighing 8 lb. But it merits attention from all lovers of good things, for no fungus yields a richer gravy, and when grilled it is scarcely to be distinguished from grilled meat.

But we must desist, our savoury theme not half discussed. The question for our readers to decide is, shall they feast upon funguses, or fast out of mere prejudice, with the despised bounties of Providence surrounding them in profusion? But there are not a few, peradventure, who will both study and eat funguses. The doing of both with advantage may be insured by carefully perusing the works now introduced to their notice. Mrs Hussey's splendid work evinces her rare powers of pictorial delineation; Berkeley is an authority in scientific mycology; while Badham, being both Reverend and M.D., is learned, witty, and jocose. Knowing the seductiveness of fungal fare, he not only prescribes medically in the case of accidental poisoning, but also preaches moderation like a Christian divine. Listen, O reader, and it shall be well with thee! "Nine-tenths of dyspeptics become so from over-feeding. Whilst it is an acknowledged fact that infants are over-fed, and that all children over-feed, men are by no means so prone or willing to admit that gluttony is perhaps the very last of childish things that they are in the habit of putting away from them. But, then, though funguses are not to be considered unwholesome, they are, like other good things, to be used with discretion, and not '*à discretion.*'"

LIFE AND HISTORY OF A SALMON.

Nature's many-leaved book is open to all her children, and in greater or in less degree we all enjoy her teaching. A child, a peasant, or a philosopher, finds something to interest him in the gambols of animals, the flight of birds, the swarming hosts of insects; the growth of trees, plants, and herbs; the properties of the air, the fantastic pageantry of the shifting clouds, the lightning's flash, and the thunder's roar.

It must be so; we cannot shut our eyes to the beauty and the beneficence by which we are surrounded. Although we remember it not, our young souls were filled with wonder when first we saw the moon walking in her brightness; and, so willing are we to make the acquaintance of any living thing, childhood has no dearer joy than to be the possessor of some favourite beast or bird. This instinctive love of natural objects, of which we know not the value or the use, does not always grow with our growth, and develop itself in the tastes and pursuits of the scientific naturalist. Our education may be, and too often is, neglected. We receive instruction only from the printed book instead of from "the infinite book of Nature's secrecy," which has charms for all, because its letters are not inky symbols, the meaning of which we painfully acquire, but the sparkling stars, the towering hills, and the heaving waves of the sea. Of late the popular mind in this country has become much more alive to the charming occupation to be found in the study of natural history. Philosophers proclaim that we cannot know ourselves, and be intelligently aware of our pre-eminence above all earthly creatures, unless we compare their material organisation with our own, and their unreasoning instincts with the capabilities of our higher intelligence. Nor is the public reluctant to listen to such philosophy. Witness the interest taken in the ontological speculations of Professor Owen, when unfolding archetypal forms, and finding the rudimentary idea of our wondrously compacted frame in the backbone of a primeval fish. Our utilitarian politicians, moreover, by the introduction of natural history into the examination of aspirants to the civil service in India, have at last publicly confessed, that not to be acquainted with the properties of material objects and with the habits of living creatures, is to deprive ourselves of the benefit resulting from that subserviency to our purposes with which they have been created, as well as expose ourselves to the injuries which many of them are capable of inflicting. We cannot afford to be ignorant of the ways and doings of any of God's creatures. They may feed us, clothe us, work for us, but they can also torment, terrify, and kill us; ravage our choicest substance, and ruin our costliest labours. A tiny insect may blast our fields, or riddle like a sieve the proudest of those floating citadels which guard our sea-girt isle.

A maritime people is peculiarly interested in the natural history of the ocean and its inhabitants, and of those great rivers and lakes which are the highways of commerce. Dryden insists that for the idea of a ship we are indebted to the form of a fish; and the sagacious Paley is confident that plate-armour was suggested by the lobster's tail.

Our interest in fishes is determined by our position.

Great Britain and Ireland have a coast-line of more than 4000 miles; and of the more than 8000 fishes described by naturalists 253 inhabit the fresh waters of Britain and the surrounding seas. Our shores abound with those kinds of fish which exist in the largest numbers, and yield a supply of the most grateful food. Our fisheries are an important branch of national industry, and add largely to our national wealth; so that a serious diminution in their produce would affect the comfort of all, and be ruinous to multitudes of our people. But while we have such special reasons for investigating the natural history of fishes, manifold are the difficulties in the way of attaining the knowledge which we seek. Dwelling in the depths of the ocean, or shunning observation in their favourite haunts in lakes and rivers, they perform, unseen for the most part, the functions of their being, so that there may be numerous species whose very forms are yet unknown. Our philosophers, moreover, as Goldsmith complained long ago, instead of studying the nature of fishes, have too often "employed themselves only in increasing their catalogues, from which all that results is but an additional tax on our memory." The amount of philosophic ignorance regarding the generation of fishes is curiously illustrated by the fanciful notions which prevailed as to that of eels. Aristotle believed that they sprang from mud; Pliny maintained that they were propagated by fragments separated from their bodies by rubbing against rocks. Helmont asserted that they came from May-dew, and gives this whimsical recipe for their production: "Cut up two turves, covered with May-dew, and lay one upon the other, the grassy sides inwards, and thus expose them to the heat of the sun: in a few hours there will be sprung from them an infinite quantity of eels!" Horse-hair from a stallion's tail, and placed in water, used to be considered an unfailing source for a supply of young eels;—a fancy which lingered till the days of our youth, for we remember trying the experiment, and with the most encouraging prospect of success, as we verily believed, when the horse-hair in the water really did begin to move!

While philosophers have thus imposed for a time on public credulity, the general ignorance regarding fishes has been greatly maintained by the prejudices of fishermen, and the incapacity of so-called "practical men" to observe accurately and reason justly on matters with which they are conversant. From the report of a parliamentary committee we learn that a tacksman of very extensive fishings actually mistook the tape-worm (a mischievous parasite infesting certain portions of the intestinal tube of the salmon) for the food of the salmon. Another "practical man" declares that the digestion of this fish is so rapid, that "fire or water could not consume quicker."

Of the ability of a fishmonger to edify the public as to the habits of the creatures which he is daily handling, we lately fell in with an amusing instance. An ancient dealer in fish, either a wag or an *ignoramus*, gravely informed the writer of "A Morning Visit to Billingsgate," recently published in 'The Leisure Hour,' a meritorious production of the London Tract Society, "that salmon-fry are developed from the egg in forty-eight hours, during which time the breeding-bed is anxiously watched by the parent fish, which, immediately on the appearance of their young, conduct them to the ocean with the tenderest care!" This, it seems, is the London fishmonger's version of the now exploded notion that salmon-fry become smolts in about there or four weeks after birth, and migrate to the sea.

It must be owned, however, that the difficulty attendant on all ichthyological researches is increased, in the case of the salmon, by its characteristic habit of annually migrating from its native river to the rich feeding-grounds of the ocean. In writing the life of a salmon we may almost be said to write two biographies; so different in habits and appearance is the salmon at sea from the salmon in the river. In the former he feeds like a glutton, in the latter he is so abstinent that food is hardly ever found in his stomach. With its ocean haunts and modes of living we are still most imperfectly acquainted; but recent discoveries and observations have removed the mystery heretofore resting on its river life, and enabled us to watch the singular process of its generation, from the extrusion of the egg from its parent's body until the little fish assumes the characteristic signs of the salmon, and takes its first departure to the sea, soon to return astonishingly grown, and become a coveted article of human food.

On account of its value—for Franklin truly described it as "a bit of silver pulled out of the water"—as well as because of its most interesting habits, we shall give our readers a brief sketch of its history, and bring under their notice certain important experiments in pisciculture, resorted to in the well-founded hope that they will ultimately lead to a vast increase in the numbers of the king of fresh-water fishes.

Although capable of living in the sea, and impelled to resort thither once a-year, the salmon is to be regarded as a fresh-water fish. It is born in a river, which it leaves for a sojourn of only a few weeks in the ocean, from whence, invigorated and ready for the propagation of its species, it returns to its native river with a certainty which has been long known and admired. Hence, when James I. of England meditated a return to his native Scotland, we find him stating that he was "drawn thereunto by a maist salmon-like affection." Looking, then, upon the salmon as a river fish, let us begin our account of its habits, with a description of its proceedings at the time and place of pairing.

Salmon prefer a rather cold climate, resorting to the streams when little above the freezing point; and the temperature in which they appear to thrive best is from 40° to 52° Fahr. Their favourite locality in Europe may be reckoned from the Moselle to the arctic circle. The spawning season varies somewhat throughout this extensive region, but in the salmon rivers of this country the variation is inconsiderable. It is believed that spawning occurs, in a few instances, in almost every month of the year; but the time for this interesting operation is chiefly in the months of November and December. Generally speaking, however, the fish are so far advanced in pregnancy as to be unfit for human food by the end of September. In the months above mentioned the salmon are in pairs, and having selected a stream with a gravelly bottom, the operation of spawning may be witnessed during the four, five, to ten days required for its completion—owing to the ova not being ready for extrusion all at the same time, or in consequence of the fish being scared and interrupted in their labours. Mr Shaw and others have maintained that the male takes no part in the toil of forming the spawning-bed. " Ephemera" and Mr Young, in their conjunct and very valuable work, 'The Book of the Salmon,' denounce all such as "feather-bed naturalists." Having a wholesome dread of such an imputation, especially from one like "Ephemera," the dictator to the sporting world who swear by 'Bell's Life in London,' we, with all humility, receive the account given by him and Mr Young of the manner in which salmon gratify their philoprogenitiveness : " A salmon-bed is constructed thus : the fish having paired, chosen their ground for bed-making, and being ready to lay in, they drop down the stream a little, and then, returning with velocity towards the spot selected, they dart their heads into the gravel, burrowing with their snouts into it. This burrowing action, assisted by the power of the fins, is performed with great force, and, the water's current aiding, the upper part or roof of the excavation is removed. The burrowing process is continued until a first nest is dug sufficiently capacious for a first deposition of ova. Then the female enters this first hollowed link of the bed, and deposits therein a portion of her ova. That done, she retires down stream, and the male instantly takes her place, and pouring, by emission, a certain quantity of milt over the deposited ova, impregnates them. After this, the fish commence a second excavation immediately above the first, and in a straight line with it. In making the excavations they relieve one another. When one fish grows tired of its work, it drops down stream until it is refreshed, and then with renovated powers resumes its labours, relieving at the same time its partner."

This account of the manner in which the ova of salmon are fecundated is no doubt the true one; and yet it is necessary to notice the speculations and experiments of certain recent writers, who maintain that the ova are impregnated while within the body of the female salmon, and by actual copulation with the male. M. Bonnet, the Genevese philosopher, writing in 1781, while modestly confessing that we know very little about the amours of fishes, positively declares, "Le mâle et la femelle sont privés des parties propres à la copulation."* Mr T. T. Stoddart, however, an enthusiastic angler and a very amusing writer, knows more about the conjugal rites of fishes, and roundly asserts that salmon are provided with copulative organs, which they undeniably use, he thinks. " I hold it," says he, "to be a palpable anomaly that no direct act of coition should be considered to take place betwixt the milter and spawner, and *that* long previous to the effusion of the ova."† Knowing Mr Stoddart to be a poet, we should have taken the liberty of passing by such a specimen of his imaginative faculty; but when Dr Robertson of Dunkeld professes to have demonstrated, by careful experiment, that the ova of trout are impregnated previous to exclusion, we feel bound to attend to the theory of our poetical angler. It is contradicted by the repeated experiments of Messrs

* 'Œuvres,' tome ix. p. 287.   † 'Angler's Companion,' p. 189.

Shaw and Young, and still more recently by those of Dr Davy, as communicated to the Royal Society of Edinburgh. All these careful observers declare that not the slightest sign of vitality could they ever perceive in the ova of salmon or of trout, unless previously impregnated by contact with the milt. As to the inaptitude of the organs of salmon for coition, Dr Davy's anatomical reasoning is conclusive.

Dr Davy also points out the probable source of Dr Robertson's mistake, when he fancied that he had succeeded in hatching trout from ova not impregnated by milt. "The box containing them was placed in a stream. What is more likely than that they might have been impregnated—so included, but not insulated —by the spermatic granules, the spermatozoa of milt shed by some fish in the adjoining water? The diffusibility of these living granules—not the least remarkable of their qualities—seems to be favourable to this conclusion."

In confirmation of Dr Davy's supposition that the ova of the trout deposited in the stream were accidentally impregnated by floating particles of spermatic fluid, we recall the singular experiment of Spallanzani in artificially fecundating the ova of toads. He found that to touch them once with the point of an extremely fine needle was enough to render them prolific. Until, therefore, the ova of trout or salmon are fertilised under circumstances rendering the access of the smallest portion of the male sperm impossible, we cannot attach importance to the theory of Mr Stoddart, or to the unsatisfactory experiments of Dr Robertson.

The ova, then, of the salmon being deposited after impregnation in the manner described, we have now to trace the process of their vivification. Of this we are not aware that we have any detailed and perfectly trustworthy account until the publication, in 1843, of 'Observations on the Natural History of the Salmon,' by Robert Knox, F.R.S.E. While cheerfully acknowledging the great value of these observations, justice requires us to record that this skilful anatomist and careful observer sanctioned the mistaken idea that, in three or four weeks after exclusion from the egg, the young salmon assumes the silvery aspect of the smolt, and migrates to the sea. For the rectification of this, until lately, general belief, we are indebted not to a man of science, but to Mr Shaw of Drumlanrig, forester to the Duke of Buccleuch.

The quantity of ova deposited by the salmon, though by no means so great as that of many fishes, is nevertheless very considerable. Those of a grilse of 8 lb. amount to about 8000, while those of a salmon of 25 lb. have been found to be nearly 25,000. The period during which they remain under the gravel varies with the temperature of the water in which they are placed. Ova deposited in September, when the temperature is high, will produce fry in 90 days; while those deposited in December may remain unproductive till the 140th day. In the artificial breeding-ponds of France the ova come to life in 60 days, whereas in Perthshire the earliest have not appeared till 120 days. A gentleman at Perth, however, having deposited ova under a spring flowing from a rock at Barnhill, the fry came into life, some in 50, and others in 60 days.

The ovum, on its extrusion from the salmon, is of a pale-blue colour, and about the size of a small pea. It is composed of a white shell and light-red yolk. On the 20th day after being deposited in the gravel, a very small but brilliant spot appears on one side of the yolk, and this brilliancy increases daily until, on the 48th day, it covers more than half the yolk, and still continues spreading. On the 48th day the young fish make their first appearance in the form of a bluish white thread in the yolk of the egg. On the 63d day this formation is furnished at one end with two very small black spots—the eyes of the future fish. After this there is a decided daily change, until the 93d day, when the eyes are more distinctly seen, and the head also is apparent; the red of the yolk is drawn more closely towards the belly of the fish, where it is formed into a conical bag, one end of which is attached to the fish. This singular appendage is better seen on the 100th day, when the motion of the embryo, and of the attached bag, may be leisurely examined, by taking an ovum, with a little water, into the hollow of the hand. The heat of the hand, raising the temperature of the water, causes immediate activity in the fish; and the whole of its structure and rapidly advancing changes may be daily watched. The restlessness of the little prisoners increases day by day until the 135th day, when the whole spawning-bed is in lively commotion, and, very many of the fish having so far made their escape as to liberate their tails, the bed resembles a thick braird of grain. The coiled-up fish breaks his cell by his efforts to straighten himself, the rupture being always effected opposite the back fin. The conical bag and the head remain in the shell a little longer; but the efforts of the fish, aided by the stream, soon effect a perfect liberation, and the newly-born fish are before us, of the average length of three quarters of an inch.

Instead of being a handsome little creature, such as we might fancy an infant salmon, it has an ungainly tadpole-like appearance, and, though rapid, is unsteady in its movements, owing to the disproportionate size of its head, and of the large heart-shaped bag still attached to its belly. Finding itself unable to swim while hampered with this appendage, fully a quarter of an inch long, and as bulky as all the rest of its body, the little creature, when disturbed, instantly darts under a stone, as a protection at once from the violence of the stream and the voracity of its enemies. We should describe this movement as a rapid wriggle rather than true swimming. It answers its purpose perfectly, however, for it is out of sight in a moment. This protuberant bag is a singular provision for supplying it with food for the first five weeks of its existence, during which period it takes no external nourishment. Its contents being consumed, it disappears; and the fringe-like fin, hitherto surrounding the fish, gradually vanishes, and is succeeded by the true fins in their proper places.

At the age of two months the fry are distinctly marked with those transverse bars which used to be considered the characteristic of a different species of fish—the parr, which, however, by the experiments of Mr Shaw, is now demonstrated to be the young of the salmon; or, to speak more correctly, the little fish generally termed parr is now shown to be the young salmon. We make this correction out of deference to those who, like Dr Knox, Mr Young, and others, maintain that there is what they term a parr-trout—an adult, as they conceive, of the river-trout species. The alleged fact of the existence of such a species of trout can only be set at rest by the production of a female with well-developed roe. Such a female is said to have been seen; but, considering " the extreme rarity of such a phenomenon," we agree with Dr Knox that " it would have been desirable to have preserved the specimen, and submitted it to scientific men." In the mean time, all that is known about parr may be found recorded in a chapter of Dr Knox's little work, ' Fish and Fishing in the Lone Glens of Scotland.' Personally we never saw, nor have we ever met any one who has seen, a female parr with well-developed roe. And Mr Ffennell, one of the inspecting commissioners of the Irish Board of Fisheries, declares that he has not been able to procure such a parr, though he has been in search of it for years, and has employed others in the search, and offered a high premium for the fish in the desired condition.

As the distinction betwixt the young salmon in the parr-like state and the so-called trout-parr is not obvious to common observation, the prudent course is to regard them as identical, and to preserve them with the utmost care. Considering that thirty years have elapsed since Mr Shaw demonstrated that what is commonly called parr is in truth a young salmon, it is marvellous that the proprietors of salmon rivers should so carefully watch over the young salmon during the two months when they chiefly assume their silver coats, but permit for ten months in the year the unrestricted slaughter of parrs, which, notwithstanding the existence of the very similar parr-trout, ought assuredly to be considered as the young of the true salmon. If our national custom was to commit infanticide, without remorse or legal hindrance, until our younkers were approaching puberty, and if the law were to denounce as murder the slaying of youth or maiden of about the age of fifteen, we should have a parallel to the piscatorial folly of permitting the slaughter of parr, but legislating stringently against the capture of smolts.

A very interesting controversy is still going on as to the precise time when the parr assumes the silvery coating of the smolt, and makes its first migration to the sea. Mr Shaw, we believe, continued to hold the opinion that this, with a few exceptions, does not occur till the conclusion of the second year. On this important point he differed with Mr Young, and also with Mr Gotlieb Boccius, who, after ample experience, speaks thus positively: "When a smolt is a year old, nature begins to cover it with a second lamination under the first scales, and thus covers the bars. It then shows a desire to migrate; but the migration to the sea cannot take place until the second formation of the scales has been perfected, by means of which new envelope its body is protected from the increased density upon the removal from fresh water to salt. It has been fully ascertained, that when fry without the second lamination, or under a year old, have been placed in salt water, they immediately die."

The determination of the time when the parr assumes

this "second lamination" became in May 1855 a question of practical importance, demanding immediate solution. At that date the salmon-fry reared in the breeding-pond at Stormontfield, near Perth, had attained the age of nearly fifteen months. Those interested in this great experiment had therefore to make up their minds as to whether they were to be dismissed from the pond, in conformity with the experiments of Mr Young, or retained for another year, as advised by Mr Shaw. The late Mr James Wilson, the well-known naturalist, Mr Shaw, along with Lord Mansfield, the late Dr Esdaile, and several others, met at Stormontfield on the 2d May 1855, in order to decide what should be done with the imprisoned fish. We give the result in the words of Mr Wilson: "The only example of a smolt exhibited to the meeting throughout our careful inquiry and investigation, was one caught by Dr Esdaile while angling in the river Tay. He brought it to us immediately; and when set alongside the parr from the pond, its greater size and spotless silvery lustre made its difference obvious to all. The meeting came to the distinct and unavoidable conclusion that the inhabitants of the pond *were still parr;* and Lord Mansfield especially, and very properly, pressed for their being detained in confinement another year, for the sake of a complete and conclusive experimental demonstration that these so-called parr take two years to become smolts. Their continued captivity was therefore determined on."

The holders of the biennial theory had thus a complete triumph. It was but of short duration, however; for, in a letter published 21st May, Dr Esdaile wrote thus: "On the 19th May, seventeen days after Mr Shaw's visit, I revisited the pond in company with a number of experienced fishermen and anglers, and out of about a dozen fish caught by me with the fly there were five unmistakable smolts, not to be distinguished from the smolts caught in the river at the same time. Many other smolts were also taken with a very imperfect net, so that there must be in the pond already many hundreds, or even thousands, of exceptions to Mr Shaw's rule, and as the weather becomes warmer, the conversion of the fish will doubtless be proportionably increased." These facts becoming known, the tacksmen of the Tay fisheries presented to the proprietors of the fishings a memorial, in which they express it as their "decided conviction that all the young fry should, on as early a day as possible, be turned out into the river, to join their fellow-smolts now on their way to the sea. But," they reasonably added, "we have no objection that a portion of the young fish be retained in the ponds to test the correctness of the views of those who maintain that they ought to remain two years, it being our anxious wish that this interesting experiment should have a fair trial, and every fair play."

In consequence of this representation, the fry in the pond were permitted to enter the river on the 29th May, care being taken to mark about 1300 by cutting off the adipose or dead fin, for the purpose of recognising them on their speedily anticipated return from the ocean.

The result shall be related when we have finished our sketch of the life of a salmon. Their progress to the sea was, it may be presumed, not unlike that of a migration thus described by Mr Shaw, from personal observation: "They passed down the river in small family groups or shoals of from forty to sixty and upwards, their rate of progression being about two miles an hour. The caution which they exercised in descending the rapids they met with in the course of their journey was very amusing. They no sooner came within the influence of any rapid current than they in an instant turned their heads up the stream, and would again and again permit themselves to be carried to the very brink, and as often retreat upwards, till at length one or two bolder than the others permitted themselves to be carried over the current, when the entire flock disappeared; and then, as soon as they had reached comparatively still water, they again turned their heads towards the sea."

Arrived at the ocean, they find that the Power which led them thither has made a bounteous provision for their wants in the eggs of various species of echinodermata, together with the smaller crustacea. Returning to the river in six or eight weeks, their increase in size is proved to be astonishing. To Dr Esdaile we are indebted for the following interesting memorandum, demonstrating this in the most satisfactory manner:—

"No. 1 is a young salmon, 15 months old, from the artificial breeding-beds and rearing-pond at Stormontfield, killed 29th May 1855; length, 5 inches; circumference over dorsal fin, 2 inches; weight, $\frac{1}{2}$ an ounce. No. 2 is a fish of the same age, dismissed from the rearing-pond on the same day, after having the dead fin cut off. It was taken by the net three miles below Perth on the 19th July, having been absent 51 days: length, $24\frac{5}{8}$ inches; circumference over dorsal fin, $12\frac{1}{2}$ inches; weight, $5\frac{1}{2}$ lb."

Next season this weight will be doubled; but, being ignorant of the extreme limits of the natural life of the salmon, we know not how long this increase may continue.

Salmon, however, have their troubles, and while revelling in the abundance of their sea-quarters make dainty fare to the nimble porpoise and the voracious seal. While dwelling in the sea they are also exposed to the minor misery of attacks by the parasitic insect popularly known as the sea-louse. They thus find that getting fat is not a sufficient compensation for being daily frightened and tormented; and, we presume, this sad experience helps their instinctive impulse to return to their native river. Their invariable return to their native streams is abundantly authenticated; the fish of each river have also such distinguishing characteristics that experienced observers will at once say, for example, "*That* is a Tay fish, while *this* is from the Isla." A fish produced in the Isla will come up the Tay, passing in its course the Erne and the Almond, and in the Isla

"Repeats the story of her birth."

Nay more, if produced in some of the streams falling into Loch Tay, it will pass we know not how many small rivers, and through the loch into the particular stream for which it seeks.

When Ovid (1. *De Ponto*) thus mourns the exile's fate, he was in truth describing salmon—

"Nescio qua natale solum dulcedine cunctos
Ducit, et immemores non sinit esse sui."

Knowing that a single salmon produces, it may be, twenty-five thousand ova, and that, according to a very moderate computation, one hundred million are annually deposited in the Tay, we are apt to be surprised at the complaints as to the decrease of salmon, not in this country only, but also in Norway, Holland, and the United States of America. But our surprise disappears when informed that the salmon annually captured in the Tay are as one to a thousand of the deposited ova. It is scarcely within the power of human ingenuity sensibly to diminish the finny tribes inhabiting the sea. And yet the whale—

"Leviathan, which God of all His works
Created hugest that swim the ocean stream"—

has by man's incessant pursuit been driven so far within the regions of "thick-ribbed ice," that whale-oil, to the great joy of gas companies, is annually becoming dearer; and ere long whalebone may be so costly that, in answer to the question, "What's the use of whalebone?" we may never again receive from a boy in a Scottish parochial school the ready reply, " Our *mithers* put it in their breasts."

But a fish alternately dwelling in the ocean and the river is so much within the reach of human destructiveness as to require artificial protection. We therefore look with interest and hope on the proceedings of those who, in Germany, Sweden, England, France, Ireland, and Scotland, are engaged in what, absurdly enough, is termed the *artificial propagation* of the salmon. This absurdity is amusingly seen on the title-page of Mr Ramsbottom's interesting pamphlet, which exhibits "a specimen of a young salmon taken from those sent to the Dublin Exhibition, and *propagated by the author!*" All that is artificial in the matter is the application of human fingers to the belly of a parturient fish, the placing of the expressed ova under gravel in a current, the confinement of the fry till the time of migration, and the feeding them on sheep's liver and occasional maggots, as at Stormontfield, or upon nothing but water, as recommended by Mr Boccius. This is possible only when the young fish are confined within a large space. At Stormontfield the size of the pond is only a quarter of an acre; and there, we are persuaded, they would have perished unless artificially fed.

For this happy idea we are indebted to the German naturalist Jacobi. Observing how flowers are impregnated by the fertilising dust being conveyed by winged insects lighting on them, it occurred to him that, in like manner, the prolific seed of one living creature could be artificially transferred to another. In 1758 he artificially impregnated trout and salmon ova. Taking the female fish when her ova were mature, he gently pressed them out into a vessel of pure water, into which, in like manner, he immediately introduced the milt of the male. Amongst the most curious of his discoveries must be reckoned his demonstration of the fact that eggs taken from a fish four or five days dead, and actually putrid, could be impregnated as successfully as those taken from a living fish. Practically applied, as Jacobi's experiments were to a considerable extent, they were apparently forgotten, when in 1849 the attention of the Government of France was attracted to the labours of two humble fishermen, Géhin and Rémy, who, by the same method practised by Jacobi, and more recently by Mr Shaw, had succeeded in stocking the tributaries of the Moselle with millions of trout; the ova, in some instances, having been transferred from distant rivers celebrated for their superior breed of fish.

Such results attracted the notice both of the capitalists and the lovers of natural history in this country. Messrs Edmund and Thomas Ashworth have, by means of artificial fecundation, reared millions of young salmon near Lough Corrib, in Ireland. Their operations trenching on the supposed rights of the squireens, we were sorry to be lately informed that these small gentry had intimated that, unless bought off, they were resolved to hinder the further prosecution of the experiment!

In the autumn of 1853, at the suggestion of Dr Esdaile, the proprietors of the salmon-fisheries on the Tay laid down 1000 lineal yards of breeding-troughs, calculated to contain a million of salmon ova. Owing to the lateness in beginning operations, the boxes were only partly filled, and in May 1855 the breeding-pond was estimated to contain about 300,000 salmon-fry. We have already alluded to the debate as to the time of their dismissal from the pond. We have now to speak of the commercial aspect of the question, and to describe a perplexing anomaly in the history of these little fish.

Of those marked we have undeniable proof that 22 were captured as grilses, weighing from 5 up to $9\frac{1}{2}$ lb. The fish of the last-mentioned weight was taken on 31st July, having left the pond on 29th May, weighing in all probability not more than a single ounce. It has thus been demonstrated that salmon fit for the market can be artificially reared within 20 months after the deposition of the ova. Of the salmon-fry first reared at Stormontfield, it was estimated that 2000 have been taken as grilses, and that, valuing them at 3s. 6d each, their worth was £300. Next year, those captured as salmon would be much more valuable; so that, as a commercial speculation, the experiment will prove remunerating, provided it be carried out on an extensive scale. The hope of the Tay-fishing proprietors, that their gain would consist of all they could capture of about 300,000 grilses, was frustrated by an unexpected occurrence. After the 7th June migration from the pond ceased, and thus there was left in it apparently one half of the fish whose departure and return the same season had been confidently reckoned on. Thus, while one portion of the same hatching was being captured in the river as beautifully-grown grilses, another portion was still in the pond, tiny creatures of about three inches long, and not more than an ounce in weight! When we visited them in August, so numerous were they, and so lively their movements, that the surface of the pond appeared to be dotted, as it were, with drops of rain. They took the fly-hook with the greatest avidity, and having captured one, we found it plump and vigorous, but a veritable parr, totally free of the silver scales of the smolt. At the end of November they exhibited the same appearances. Mr Buist, the intelligent superintendent of the Tay fishings, assures us that "no other fish, nor even the seed of them, could by possibility get into the ponds;" and he also mentions the singular fact, that in these little parr "the males have the milt as much developed, in proportion to the size of the fish, as their brethren of the same age, 7 to 10 lb. weight; while the females have their ova so undeveloped that the granulations can scarcely be discovered by a lens of some power."

It is evident that the opposing views of Messrs Shaw and Young are partly right and partly wrong. It is not true that only a small portion of salmon-fry become smolts at the end of their first year, as Mr Shaw supposes; neither do they all become so, as Mr Young confidently asserts.

From Mr Shaw's admirable memoir "On the Growth and Migration of the Sea-Trout of the Solway,"* we learn that, at the age of twenty-four months, three-fourths of the brood of that fish assumed the migratory dress, and that this was not assumed by the remaining fourth, which he maintains do not migrate, but permanently reside in the river; the females maturing their roe sufficiently to reproduce their species with young males of corresponding age. A writer in 'The Edinburgh Review' (July 1843), having referred to the sea-trout permanently resident in the island of Lismore, Argyleshire, thus proceeds: "We venture to state that this same power of adaptation to fresh water is possessed by the salmon;" and having referred to Lloyd's 'Field-Sports in the North of Europe,' to prove that the 21,817 salmon caught in Lake Wenern in 1820, of the average weight of six or seven pounds, but never above twenty, "could never have been in the sea," the reviewer proceeds: "It is probable, indeed—and we beg to call Mr Shaw's attention to the probability—that a certain portion of the salmon-brood may, like the sea-trout, remain permanently in our rivers."

It remains to be proved that these so-called salmon in Lake Wenern are true salmon. The probability suggested may, by the experiment at Stormontfield, turn out to be a reality; and in that event, the question will arise whether the non-migration of the fish there reared be natural to the species, or the result of the circumstances in which they have been placed. It is conceivable that a young salmon, regularly fed, may have its craving for food so abundantly gratified as not to feel the instinctive longing which impels its unpampered congeners to make for the feeding-grounds of the ocean. Such a result may have important consequences. If fishes, naturally migratory, can be so treated as to convert them into permanent dwellers in our lakes and

---

* 'Transactions of Royal Society of Edinburgh,' vol. xv. part iii. p. 369.

rivers, we shall be great gainers by restraining their vagrancy. If we place them in accessible positions, where at any time we can lay our hands upon them— if we secure them from the attacks of otters, seals, and porpoises—we may, in almost every country, have lakes like Lake Wenern, yielding thousands of valuable salmonidæ, if not true salmon. The Irish Board of Fisheries transferred artificially-reared salmon-fry to a sea-pond at Kingstown, in the hope of raising them to maturity. They increased in size, but gradually disappeared, having been devoured by crabs, conger-eels, or water-rats, with which the place was infested. Mr Ffennel informs us that he will repeat the experiment under more favourable circumstances, and is confident as to its success.

In Scotland we have about 180 miles of canals. Why are they fishless, when they might be easily stocked with many valuable species of fish? Why do the Water Company, owners of the Compensation Pond among the Pentland Hills, not try to swell their annual dividend by the introduction into their capacious reservoirs of the species of salmon so abundant in Lake Wenern? In order to induce a private proprietor to do what a water company may not have the spirit to attempt, we suggest a locality admirably suited for the experiment.

If our readers have ever travelled by rail from Edinburgh to Perth viâ Fife, they doubtless remember the little Loch of Lindores, when approaching the estuary of the Tay, and at the distance of two or three miles from Newburgh. Within half an hour's journey from the populous towns of Perth and Dundee, and distant from Edinburgh and Glasgow not more than two and four hours respectively, such a locality has every advantage for carrying out the experiment we suggest. As a result of attention to pisciculture, it may be reasonably hoped that our rivers shall teem with the noble salmon; that our lakes and streams shall be stocked with the choicest kinds of fish, whose ova can be transported from foreign waters; and that grateful additions shall be made to the limited fare of the great body of the people. In those anticipated days, when fish shall be reared by millions, how many a keen angler, who now only gets "glorious nibbles," shall experience the frequent joy of slaying a plump five-and-twenty pound salmon fresh from the sea, with the sea-louse yet clinging to his silvery sides! And how shall his heart beat against his ribs as the noble fish springs yard high into the air, and then dashes madly down the stream, while the reel *birrs* as the line whisks through his almost blistered fingers! If in the plenitude of their piscatorial skill some of our readers have only "dragged an incautious minnow from the brook," they will be amazed to be told that the first capture of a bouncing salmon is an era in the angler's life, and lets him know what stuff he is made of. Why, we have seen a fellow six feet high trembling all the length of him, when first called upon to deal with all the cunning dodges, the hard tugs, the wild springs of a well-hooked heavy salmon. The bodily labour of a prolonged contest with a powerful fish is far from inconsiderable. To follow him for a couple of hours, perhaps, along the wild banks of a roaring river, sometimes running, often wading, and it may be stumbling over the smooth stones, is exercise calling into vigorous play the muscles of the back, arms, and legs, and affording such excitement to the mind, moreover, that we do not wonder at the enthusiastic declaration of Mr Younger, the well-known angling shoemaker of St Boswell's: " It beats Grecian games, as well as English horse-racing and hound-coursing, all to nonsense: for of all earthly recreations, that of a start for a day along a fine trouting stream, by grassy bank and alder copse, with the excitement of having something to pursue as an object of exercise, has a perfect charm in it, giving a refreshing relish to the existence of the recluse of art, or the son of craft, shut up the year long in the stalls of labour, where even a rat, though fed to the full, would tire, and eat his way out through a deal board!" We look upon outdoor recreations for the labouring classes as deserving of all encouragement; and when streams and canals, lakes and rivers, come to be amply supplied with variety of choice fish, we hope that proprietors of fishings will throw them open to all rod-fishers every Saturday throughout the season, in compensation for the rigorous preservation of the parr. The people's natural love of sport will thus be gratified without incurring the stigma of poaching; and all classes having an interest in the preservation of the fish, they will be allowed to breed unmolested. It will be reckoned shameful to kill them out of season; and, public opinion being on the side of the law, no mercy will be shown to the poacher who spears or nets salmon on the spawning-bed; so that never again shall we hear of a merciless slayer of mother fishes confessing before a parliamentary committee that, at one hauling-place on the Tweed, his brother one night netted upwards of 400 spawning salmon and grilses; which, we add for the gratification of our Scottish friends, were sent to Edinburgh as *kippers*,— the red colour, which looked so charming, arising from the liberal use of saltpetre!

SALMON AND PISCICULTURE.

The popular mind in this country has of late become much more alive to the charming occupation to be found in the study of natural history. We find a new proof of this in the contents of a number of the 'Quarterly Review.' There are no less than three articles, respectively entitled "Ferns and their Portraits," "Rats," "The Salmon." The last of these attracted our attention, as well as that of a learned *pisciculturist*, who thus wrote to us: "A good article should, I take it, resemble a good trouting stream—be rapid and sparkling, but never shallow, and have occasional pools of thought." We are quite willing to admit that these, to a considerable extent, are the characteristics of the article in the 'Quarterly.' We hope, however, that this "Triton among the minnows" will not swallow up us small-fry for presuming to hint that he is fairly among the *shallows* when thus *rapidly* arriving at the conclusion that *pisciculture* will not pay. "The peculiarity of *pisciculture* as applied to salmon is, that, as soon as you have brought your progeny past the perils of birth and infancy, you must let them forth to the world of waters *without the millionth part of a chance that they will ever return to reward their early benefactors.*" Are naturalists, then, we ask, romancing when they describe the unfailing return of the salmon to its native river?

STATISTICS OF PISCICULTURE.

We can give no quarter to such piscine heresy; we stick to what we deem the orthodox and ofttimes-demonstrated fact that salmon-fry, leaving a river in April or May, will certainly return in myriads in about from fifty to sixty days, and so astonishingly grown as to make them a coveted article of human food. We repeat that the result of the experiment at Stormontfield has proved that "salmon fit for the market can be artificially reared within twenty months after the deposition of the ova:" and that "as a commercial speculation the experiment will prove remunerating, provided it be carried out on an extensive scale."

But the extent to which the experiment has been carried in Perthshire is sufficient to demonstrate that there is not a vague hope, but a certainty, of catching a very considerable proportion of salmon artificially reared. Before the result of that experiment was ascertained, we remember having a conversation with Mr Buist, Conservator of the Tay fishings, as to the probable amount of salmon-fry returning to the river as grilses. At that time there were no *data* bearing upon this point; but we were able to inspire Mr Buist with hope, by reminding him of Mr Shaw's experiment on the fry of the sea-trout. He marked 524, by cutting off the adipose fin, in the summer of 1834. Next summer he recaptured sixty-eight of them, of an average weight of 2½ lb. On these he put a second distinctive mark, and returned them to the river. Next summer (1836) he recaptured about one in twenty of these, averaging 4 lb. weight. Marking them for the third time, he returned them to the river. On 23d August of the subsequent summer (1837), he recaptured one of them weighing 6 lb. So far, then, from there being "not the millionth part of a chance that they would ever return to reward their early benefactors," these sea-trout actually returned thrice to the Nith, their native river, and, on their first return, were captured in the proportion of more than a seventh of their entire num-

F

ber! The heart of the Conservator of the Tay fishings was gladdened by the hope we held out to him that *his* fishy progeny would be equally sure to return within reach of his murderous devices! We rejoice that it has been fulfilled, as we learn from the "statistics of the Stormontfield Pond," with which he has obligingly furnished us. The following are the principal facts: —" 1st, Of the marked fish liberated from the pond, four per cent were recaptured either as grilse or salmon. 2d, More than 300,000 were artificially reared and liberated; forty out of every thousand were recaptured; and as 300,000 were liberated, it follows that 12,000 of the salmon taken in the Tay were pond-bred fish. 3d, The average annual capture of Tay salmon and grilse is 70,000. This is *rather* better than the 'Quarterly's' dismal prognostic of the millionth part of a chance! The rental of the Tay fishings has risen ten per cent!

Bearing in mind that the fishings at Stormontfield were for a year kept from the dangers which, doubtless, far more than decimated their contemporaries in the open river, and that during all this time they were receiving abundant supplies from the careful hands of " Peter of the Pools," it is not surprising that we should have to record results so satisfactory. And that the increase in the produce of the Tay fishings is rightly attributed to pisciculture seems to be proved. There has been no similar increase in the produce of any other Scottish river. The salmon-fishing last year was deficient universally, with the exception of the Tay.

We have, therefore, the utmost reason to regard *pisciculture* as a certain method of speedily replenishing our, alas! too many almost fishless rivers.

A friend of ours returning from India forthwith repaired to his native stream, to renew, as he fondly hoped, the piscatorial exploits of his youth. He toiled all day, but caught nothing save a cruel pain in the back. Good, however, comes out of evil. Disgusted, he vowed to fish no more. Taking a last look at rod and tackle, he turned to his books for consolation. He found it in a French treatise on pisciculture. Fired with a vision of fins, he propagated the new idea, till at last it assumed a local habitation in the now famous salmon-rearing pond at Stormontfield.

Nature is no doubt a bountiful mother, but her system, notwithstanding, is founded on the principle that the larger creatures shall devour the smaller, and that all creatures in the first stage of their existence shall be exposed to innumerable accidents, tending very sensibly to the limitation of their numbers. The salmon is not only exposed to these natural causes of destruction, but also to the *malice prepense* of all-devouring man, whose wiles are so successful that recourse must be had to artificial rearing, unless we desire the almost total extirpation of this noble fish. We therefore take a lively interest in pisciculture, and have endeavoured to enlist the public feeling in its favour by detailing the result of its introduction into Perthshire.

Considering how gratifying this has been to the fishing proprietors of the Tay, we are astonished at the small scale of their doings at Stormontfield. There is only one feeding-pond, of a quarter of an acre in size. A second is indispensable; because, as about half of the fry refuse to leave it at the end of the first year, newly-produced fry cannot be introduced into the pond without the certainty of being devoured by their elder brethren. From the want of a second pond, no ova have been deposited in the breeding-boxes during last winter. And so a whole salmon-producing year has been lost for the sake of a small economy. Surely this is penny wise and pound foolish. As Dr Esdaile pointed out (in his printed letter to the Tay proprietors, 27th August 1853), the Stormontfield lade furnishes unrivalled facilities for rearing young salmon, without having recourse to an apparatus of breeding-troughs. If these are not taken advantage of, and if the successful result of the artificial rearing of young salmon do not lead to a great extension of their operations, we shall take the liberty of saying to the Tay proprietors that their wisdom is akin to that of the man who killed the goose that laid the golden eggs.

But as few are proprietors of a salmon river, we desire to impress upon all living near canal, lake, or stream, that, by transporting to it the ova of fish, any body of water may be stocked with valuable kinds of fish. This has been done on a most extensive scale in many parts of France, where fish are annually reared by millions. In addition to its commercial importance as a new branch of industry, and its social value as supplying a vast addition to the food of the people, this system opens a boundless field to scientific curiosity. In the Danube and the Rhine, the Elbe and the Spree, and almost every other river in Germany—in the rivers and lakes of Russia and Northern Europe—in the lakes of Switzerland—in the rivers of France—there exist either species of fish which we do not possess, or varieties of species which we are not acquainted with. There is every reason for believing that very many, if not all, of them may be naturalised in our waters. We do not think there is exaggeration in the opinion of M. Isidore Geoffroy Saint-Hilaire, the learned Professor of Zoology to the Faculty of Sciences at Paris:—" Pisciculture is to our waters what agriculture is to our soil, and is also called upon to add largely to our alimentary resources."

For the information of those interested in the rearing of salmon, and perhaps anxious to have a sketch of the salmon-rearing arrangements at Stormontfield, we remark that such a cumbrous apparatus is unnecessary. That patented by Mr Boccius, and containing 25,000 salmon ova, is only 2 feet long by 1 broad, and requires 4 inches depth of water. To obviate the inconveniences of breeding-boxes, which are not only expensive, but apt to collect insects which devour and mud which smothers the ova, Dr Esdaile suggests dispensing both with boxes and gravel, and depositing each egg in a little cup punched in either a plate of zinc or in a slab of coarse stoneware. Why not of strong coarse glass, which is now cheap, and in which small hollows could easily be made for the separate reception of each ovum? In about three feet square he calculates on depositing 20,000 ova, at a very trifling expense. In order that our readers may have a choice of fish-breeding apparatus, we advise them to examine that used by M. Coste, whose treatise on pisciculture, translated by M. De Valmont, and published by the enterprising Mr Thomas Ashworth, may be had of Simpkin, Marshall, and Co., London.

The public interest in the artificial rearing of salmon will, we hope, be increased by the process being witnessed in that wonder of wonders, the Crystal Palace. Being at Sydenham on a visit to a piscicultural friend, we joined in persuading one of the Crystal Palace directors that a salmon-breeding apparatus would be an attractive addition to the instructive contents of the Palace. Our friend, moreover, volunteering to supply the ova from Tay salmon, the offer was accepted; and in the Crystal Palace there soon was a model river, stocked with ova of salmon in process of being developed. Contrary to our friend's wishes, it was placed in the tropical department. As might have been foreseen, the excessive heat was most injurious. Instead of being in a temperature of about 40° Fahr., they had to endure one of about 60° Fahr. The temperature of fishes is from 2.7° to 3.6° higher than that of the medium in which they live;* so that we were not

* Liebig's 'Letters on Chemistry,' p. 67. In order to prevent unnecessary pains to protect salmon ova from the effects of cold, it should be known that, if gradually applied, it does not injure them, and that they have been known to be frozen up in a sheet of ice without losing their vitality. Réaumur demonstrated that the eggs of many insects are equally uninjured by excessive cold.

surprised to learn that the mortality among the ova was great. The gentleman in Perthshire, who obligingly supplied them, is in wrath, declares the Sydenham naturalists "naturals," and wonders that they did not "put them below a clockin' hen!" A facetious friend observes that these unfortunate young salmon are parrboiled (parboiled?)

We notice this mishap as a warning to all pisciculturists. Artificial *propagation* of fish is a misnomer, for nature will not be tampered with. Artificial *rearing* of fish is possible and profitable only by carefully studying their natural habits. With this exception, we have heard of no blunder in the management of salmon ova deposited in breeding-troughs.

We have tried the patience of non-piscatory readers, doubtless;—but, as Christopher North has said, "We love all kinds of fishing, from the minnow to the whale." We remember, too, Burton's saying—"If so be the angler catch no fish, yet hath he a wholesome walk to the brook and pleasant shade by the sweet silver streams,"—where, amid their chosen haunts, he sees the finny tribes leading, as Leigh Hunt has sung,

"A cold, sweet, silver life, wrapped in round waves,
   Quickened with touches of transporting fear."

Being benevolently desirous that the poor should have fish to eat, that the scientific should have fish to observe, and that the angler should have fish to catch, we hope our readers will pardon us for having detained them so long contemplating the ways and doings of men and fishes in fulfilling their respective lot—to eat and to be eaten.

We have a concluding request to make to those practically interested in the artificial rearing of fish. We desiderate from them such statistics as we possess regarding Stormontfield salmon and the sea-trout of the Nith. Pisciculture has been introduced in various localities in Scotland. What has been the result?

what the cost? Have the puzzling anomalies exhibited by the Stormontfield fry been witnessed elsewhere? We observe that an angling club in the west of Scotland has introduced the grayling from England. We should like to know about the introduction among us of any new species of fish, and earnestly commend this hitherto neglected field of enterprise to the attention of all desirous to cater for the public stomach, in the well-founded expectation of thereby replenishing their own purses. To all such we emphatically declare that fish-rearing, if followed out extensively, will infallibly *Pay*.

SCOTCH SALMON AND SCOTCH LAW.*

BEING lately at an Edinburgh dinner-party, abundantly supplied with all the delicacies of the season, we were struck by the very decided preference for roast mutton evinced by a gentleman from the country. In answer to the invitation of the courteous hostess to "take a little bit of this fine turkey," our friend observed —"Poultry is no rarity with me; but mutton is so scarce and so dear in my neighbourhood, that 'ere long I hardly expect to meet with it. Another slice, if you please, sir, in case that be the last leg of mutton I may see for a while."

We have learned by doleful experience that there is reason for our friend's anticipation as to the increased price of butcher-meat. Three weeks ago, in a certain county town, beef and mutton were selling at 8d. per lb. The price is now 10d.; and our butcher threatens speedily to raise it to a shilling—a calamity to housekeepers evidently at our door, seeing that this price is already demanded in a neighbouring town. If the failing produce of the land impel us to the cultivation of those never-failing harvests furnished by the ocean, and

* 'View of the Salmon Fishery of Scotland. With Observations on the Nature, Habits, and Instincts of the Salmon; and on the Law as affecting the Rights of Parties,' &c. By the late Murdo Mackenzie, Esq. of Ardross and Dundonnell. With Appendix. William Blackwood & Sons, Edinburgh and London. 1860.

which, to a more limited yet most appreciable extent, may also be found in our numerous lakes and rivers; if diminished ability, day by day, to fill our flesh-pots with the wonted supply of butcher-meat send the public, with ravenous appetite, to the perusal of our piscicultural papers, we shall benevolently rejoice. And the public, that omnivorous monster with its millions of stomachs, shall share in our joy, and be comforted by knowing that though the glories of Smithfield should depart in consequence of neither bullock, sheep, nor swine being despatched thither from any part of Scotland; yet London need not pine with hunger, nor famine look out of the eyes of those dwelling in "the granite city," because of the disappearance of the proverbial "cauld kail in Aberdeen."

Seriously speaking, we shall reckon it a matter for hearty congratulation if the unusually high price of butcher-meat shall make our Scottish people more familiar with the excellency of a fish diet. The circumstances of the present rise in the price of beef and mutton, and of the unusually abundant supply of fish, may for a time lead to a greater consumption of the latter species of food. But, we fear, we shall soon relapse into our old non-fish-eating ways, and continue to exhibit to the world the singular phenomenon of a great maritime nation, possessors of numerous and valuable fisheries, depressed by vast numbers of the population living only a few degrees above starvation, because they have not been stimulated to put forth their hands and avail themselves of the easily-attained aliment stored up in the bounteous reservoirs of the earth-encircling sea. The rich take an interest in salmon as a dainty for the table, or as affording sport during their residence in the country; commercial companies have persecuted the whale to the farthest regions of the polar sea; cod and herring innumerable are constantly being circumvented by foul means as well as fair—still the fact remains that the resources of our fisheries are very imperfectly developed; that the laws which regulate them are often absurd, and endless causes of ruinous litigation; and that in this country there is, perhaps, no department of human industry more in need of enlightened legislation than that connected with the British Fisheries. Of all this, and especially of the unsatisfactory state of the law, the work of the late Mr Mackenzie of Ardross and Dundonnell supplies many remarkable illustrations. The work altogether is something out of the common, both from the circumstances which called it forth, and from, to use a favourite phrase of the author, its *piquant* style. Mr Mackenzie, proprietor of the river Shinn in Sutherlandshire, and tenant or tacksman of nearly the whole of the four salmon rivers which combine to form the Frith of Dornoch, was forced to raise an action against Houston, possessor of some land called Creich, in the estuary, for intercepting the salmon proceeding to the rivers, at a part where the channel of the estuary, at low water, was one-third less in breadth than the Thames at Westminster Bridge, without having either a crown grant, or, it is alleged, any right whatever, authorising him so to do. The singular result was, that the Court of Session, on precisely the same point, gave two judgments directly in the teeth of each other. The latter of these being against Mr Mackenzie, he appealed to the House of Lords, which decided against him on the technical ground that he had averred no right of property *at the spot* where Houston fished; no heed being apparently given to the strong point in his case,—namely, that the fish, intercepted by Houston at this particular spot, were all on their way up to the waters possessed by the appellant. As this judgment was given in 1831, and as a considerable portion of the book is occupied with very caustic and intelligent comments on Mr Kennedy's Committee, which reported on the Salmon Fisheries of Scotland so far back as 1824, it may be thought that the present proprietor of Ardross might as well have not published, at this time of day, the opinions of his late father, expressed, as they always are, with the utmost frankness, and very often with the most thorough contempt for lawyers. We, of course, read the book with the allowance reasonably due to the fact of its being the production of a disappointed litigant; we laugh at its lively sallies at the expense of interested witnesses, making out a case in favour of stake-net fishing; we admire the grimly facetious civility with which the Laird of Ardross thinks it decent to speak of the Judges of the Court of Session, whose acquaintance with fish, and the ancient statutes thereanent, he holds so cheap. Still, though the history of the salmon be now more correctly known, there is substantial stuff in the book worthy of the attention of all interested in the salmon fisheries, and we like it all the better for its fresh *piquant* style. There is evidently sport in store for us when we find the chapter on Mr Kennedy's Committee having for its motto certain lines from the French, intimating that "the world is full of asses, and that, if we would not see them, we must get into a hole and break our looking-glasses."* The prospect of something racy thus held out is abundantly verified. The late Laird of Ardross modestly professes that an *oar* would suit his hand much better than a pen. That it would have been dangerous for any man to presume on Mr Mackenzie's inability to bespatter an opponent with well-aimed printer's ink is proved by the vigour with which he assails every witness on the stake-net side of the question. For instance, the Committee ask Mr Halliday, "Are there a great many salmon which come into friths that do not go to the rivers, but return again to the sea?" "There are a great many." "Do you mean that they are going down *from their own natural*

---

* " Le monde est plein des fous,
 Et qui veut n'en voir,
 Doit se nicher dans un trou
 Et casser son miroir."

impulse, or from being carried down with the tide?" "They are going down *from their own natural impulse* with the tide." Whereupon our wrathful author thus falls foul both of the witness and his questioner: "Had Mr Kennedy known anything of the principles of the migratory system, he could never have supposed that, after the salmon had left their migratory abode in the ocean, and reached the mouths of the rivers, they would, without any apparent cause, turn back again, any more than a flock of lapwings or woodcocks would do so after having reached our shores, and would not have asked an ignorant man such ridiculous questions; and which, besides, common sense might have shown him it was *impossible* the witness could answer,—for what means could the witness, or any man, have of discovering whether the fish went down *from their own natural impulse* with the tide, or, in short, whether they went down at all, since he could not see their movements under water even in the space of 100 yards, much less over the whole expanse of the Frith?"

The late Rev. Dr Fleming, though admitted to be a naturalist and a man of science, and therefore a witness of a higher caste, is treated with equally little ceremony: "How far" (the Committee ask) "do you think the salmon may go into a river, when not obstructed, and return to the sea without having spawned?" The doctor, having previously stated that salmon frequently enter rivers in order to escape from the numerous foes which persecute them in the estuaries, observed—"A great deal must depend on the *degree* of terror of the fish, its strength, and the state of the river." Our critic is down upon the man of science in a trice, and declares that the doctor in divinity is shockingly heterodox in supposing that the migratory instinct of the salmon is inadequate for the purpose for which it was intended; "which, in other words, is to deny the principle of perfection in the works of the Deity. This we would not expect from a son of the Church." Certain witnesses having declared that salmon enter rivers to get rid of sea-lice, and in search of food, Mr Mackenzie thus proceeds—" The doctor has discovered another reason—viz. terror. This is a *new* reason why salmon enter rivers, which certainly has no connection with any of what appear to be the principles of the migratory system. The degree of terror in the Tay fish must be very great, for many of them never stop till they reach Loch Tay. We wonder how many *degrees* of terror salmon experience, or how long they continue. We always thought that the eye was the principal organ of terror in a salmon, and that in this point the adage ' out of sight out of mind ' might be applied to him; but the doctor makes the length of his residence in a river to depend upon the strength of his *memory*. What excellent memories, therefore, salmon must have! for Mr Stephen has furnished proof that, after they enter rivers, they never leave them till they have spawned."

Fortunately for the salmon, his sense of pain does not seem to be acute, and his memory cannot be retentive, for salmon which broke away part of the angler's tackle, and so escaped, have often been taken shortly afterwards with the identical tackle in their jaws; and Mr Mackenzie declares that he knew a case in which a salmon, having lost an eye on the hook, made use of the remaining eye to *rise* at the same hook, by which it was caught on the same day!

But the witness against whom our author directs the main battery of his ridicule is a Mr Stevenson, who deponed with all gravity to this most gullible Committee —" The nearer salmon are got to salt water the finer is their quality; so much so, that any one versed in the state of salmon, would at once be able to pick out from five hundred head of fish those that had been more than two or three days in the river; indeed, I am not sure that I could not distinguish the fish which had been taken *one mile* from the sea!" Now, a salmon will run a mile in less than ten minutes or a quarter of an hour.

What an admirable *coup-d'œil* Mr Stevenson must have —*qu'il a le nez fin!* Baron Munchausen himself could scarcely do more. We wonder how Mr Stevenson discovered *ce beau secret*. We suspect he must have found it from the old rascal of a seal who misled Mr Halliday —for the Highlanders believe that the fallen spirits have been sent into seals; and certainly it required something more than human—or, as our neighbours would say, *il fallait être un peu plus que le diable* *—to have discovered it."

This unfortunate witness having further declared that *there cannot be a doubt* that salmon *spawn in the sea*, his merciless tormentor asks him, "Did the seal tell him this also? Perhaps, being a great sportsman, he has met with some wild-ducks that build their nests in trees, while others make their nests on the ground, and has sagely concluded from thence that some salmon spawn in fresh waters, and others in the sea!" This unmitigated nonsense is only paralleled by that of one of the luminaries of the Scottish bar, who declared that if all the rivers in the kingdom were blocked up, salmon would become more plentiful than ever, as they would then be *forced* to spawn in the sea! Only fancy such a man on the bench deciding a case involving acquaintance with the natural history of fishes of the salmon kind! We have referred to Mr Mackenzie's sagacious and witty comments on the evidence laid before Mr Kennedy's Committee—not for the sake of amusing our readers with his lively sallies. The Blue-Book containing it is, in fact, the grand repository in which are preserved the facts and arguments relied upon by the partisans of the stake-net fishing interest; and these are of greater interest now than ever, in consequence of the decision of the House of Lords in the recent case of Gammell *v.* the Commissioners of Woods and Forests. It has been decided that the Crown is proprietor of all salmon-fishings on the coast of Scotland, except so far

* One would need to be a little more knowing than the devil.

as they had been granted out to the Crown vassals by charter. This decision, it may easily be conceived, has excited much trepidation along all our shores, and has caused many a search after the now indispensable charter. It is also reasonably inferred that the Crown will not permit this noble heritage to lie waste, but that it will stimulate salmon-fishing on the coast, by granting charters for a consideration, the amount of which may largely swell the public revenues. Supposing this course to be followed, the question of stake-net fishing must be reconsidered in all its varied bearings upon the rights of individuals, and upon the interests of the public. Is it legal? Judging from the celebrated Tay case, about which there was such fierce contention for twenty years, many a lawyer will feather his nest before that question be judicially answered. And then the contradictory statements of those professing to be familiar with the natural history of the salmon! Some, like our author, aver most truly, we think, that he is a river fish, migratory and gregarious; a true Highlander, born amid the mountains, but who, like other Highlanders, goes to forage elsewhere. Others, like Dr Fleming, insist that he is a sea fish, because, during the period of his migration, he is a dweller in the ocean. On this principle a youth born in Scotland, but going to push his fortune in the wider and more fertile field of "merry England," which he forsook as soon as he had made himself "comfortable," should be reckoned an Englishman.

But the fiercest conflict will be between the opposing interests of the fishing proprietors on the coast and the proprietors of salmon rivers. It must be owned that a maritime proprietor, with the sea at his door as his huge fish-pond, is grievously tempted to avail himself of the stake-net as the readiest method of capturing multitudes of coveted salmon. But the natural history of the fish, and the rights of river proprietors, demand that the owners of maritime fishings shall not be permitted to erect all manner of destructive engines for the purpose of intercepting salmon on their return to their native waters, where alone they can propagate. It is here that the interest of the public comes in. The national stomach desiderates salmon, and this all the more eagerly the rarer it becomes. The national conscience has not patience to expiscate the facts in a twenty years' litigation between the caterers for the public maw. The cry is, "Give us salmon! abundantly and cheaply, if possible; if not, at any price; by what means we care not. Let us have salmon caught by net and coble, by stake or stage, still, poke, or bag-net, we inquire not which." Oh guzzling and gullible public! why speak thus senselessly? Is the fable of the fool who, covetous of money in hand, killed the goose that laid the golden eggs, clean forgotten? Will you not listen to the "fresh-water proprietors," when, in answer to the charge of being greedy monopolists, they undertake to demonstrate that rivers are the true source of the salmon-fishery; that to hinder the salmon having free access to them is a fatal barrier to the prosperity of salmon fisheries, and a grievous damage to the national wealth; that stake-net fishing cannot possibly add to the number of salmon sent to market, but necessarily tends to its rapid diminution; and that when most successfully followed, the market price of salmon did not fall one farthing; in short, that all sorts of fixed engines for the capture of salmon in the sea are injurious to the common weal, and unjust to river proprietors, whose waters are the true nurseries of the salmon? To read this long sentence will take away the breath of the thoughtless public, and before it is finished we think we hear a clearing of the throat—an audible hem, at last finding speech in articulate words. "Oh! we did not think of all that; we did not know that the more salmon we kill in the sea this year, the fewer will breed in the rivers, and consequently the fewer will return to the sea next year, and so the fewer can be caught in stake-nets." But the public must submit to be told that, unless people will shake off sloth, and inquire into facts and weigh evidence, they shall offend against the laws which preserve the existence of all living creatures, and shall bring upon themselves needless poverty and loss. The public has a manifest right to intervene in the squabble between the belligerents striving for the possession of the salmon: law, justice, and the public interest demand that this strife shall be equitably adjusted. We confess that we see no injustice in saying to proprietors on the coast,— Use only the net and coble in the capture of salmon; or, at least, agree to employ no fixed apparatus within five miles of the mouth of any salmon river. If they make wry faces at this reasonable proposal, we will not, as they would think, mock them by advising them to try fly-fishing for salmon in the sea, though we have heard of it as most successful. We remind them that their right to do with their own what they please is— in the matter of salmon—not so clear as they fancy, and that the ancient laws of Scotland regarding salmon-fishing may be invoked against them. We shall, with the help of Mr Mackenzie's amusing chapters on "Rights of Parties" and "Scottish Statutes," endeavour to make our readers acquainted with what may be said in favour of the fresh-water proprietors. And in order that their along-shore antagonists may not suspect us of prejudice generated by self-interest, we hereby depone that we possess neither land nor water in fee-simple, and that our interest in salmon is merely that of a friend of the persecuted finny tribes, and a student of questions affecting social economy. If we seem to bear hard on maritime proprietors, it is not because, like Justice, we are blind, but because we have eyes accustomed to look around with freedom, and are thus aware of the multiform aspects under which the debatable matter of the fisheries presents itself to the impartial observer. If desirous to form an unprejudiced opinion regarding what is for the public good, we should not listen only to one side of the question. A wrathful "fresh-water proprietor" is not a safe guide, any more than a stake-net owner, who, planting his murderous engine at the mouth of a salmon river, professes his ability to supply us with salmon to our heart's content. We are in more trustworthy hands when availing ourselves of the ample experience of the Duke of Richmond. His Grace being proprietor both of sea and river fishings, and having been induced to try which mode of fishing is most remunerative, has deliberately removed stake and bag nets, with two exceptions. We state the result as given by his Grace, when under examination as a witness on the Tweed Fisheries Bill.

"I am the proprietor of salmon-fisheries in the Spey. They extend about nine miles and a half. I have also estates upon the breeding-grounds. I have likewise several miles of shore-fishings near the mouth of the Spey. They were let in 1848 for seven years; but the tenant asked to be relieved from the lease, and I complied with his request. I suppose he did not find them answer his purpose. I advertised them in the 'Times' and other papers, but nobody would take them. Prior to that they had been rather severely fished. There were sixteen stake and bag nets east and west the mouth of the river. I have still a stake-net at Port-Gordon, and another elsewhere, but all the others have been removed. My object in taking them away was to allow the fish to enter the river from the sea, because I had found that the nets frightened them when they attempted to enter the river, and drove them again to sea. I have paid attention to the breeding of salmon. The breeding depends upon the existence of fish in the river, and the protection given to smolts. You cannot have any quantity of fish in a river unless you allow them to go up from the sea; and if it can be shown that the few stake-nets which still exist are injurious in that respect, I shall remove them at once." The result of this experiment has been perfectly satisfactory. In 1851, the fish caught were 48,460, and in 1858, 81,562. In 1857 the captures amounted to no less than 99,888. The rental of the Spey fisheries was £6000 per annum, whereas his Grace's profit in 1858, when the fisheries were in his own hands, and after he had removed the stake and bag nets, and lengthened the annual close-time, was £12,445.

This experiment is of the highest interest. A nobleman, fortunately the possessor alike of coast and river fishings, declares that his almost total suppression of stake and bag nets leads to an immense increase in the number of salmon, and to more than double the former profit. He is driven to occupy his own fisheries, because he cannot find a tenant while such destructive modes of fishing are permitted. If it be said that his Grace might by stake-nets have caught as many salmon at sea as were taken in the river, and that the supply to the public would have been the same, we demur to that statement. The salmon would have been frightened out to sea, to be the prey of porpoises and seals, instead of being allowed to re-enter their native river, there to exercise their generative functions. And if not allowed to generate, we cannot wonder that they should diminish in numbers. Wherever stake-nets are introduced, the value of river-fishings is forthwith lowered, and in proportion to the increase of captures by means of such nets. During the last three years of stake-net fishing in the estuary of the Tay—viz. 1810-11-12—the average annual export from the river-fishings was only 1665 boxes of fish; while, according to the statement of the stake-net fishers themselves, their exports amounted to 4000 boxes, making together 5565 boxes yearly. But during the first three years after stake-nets were removed, the average annual amount of the river-fisheries was 4552 boxes; and during the next three years, after the river had somewhat recovered from the effects of the stake-net system, it amounted to 5930 boxes, being, during these three years, 265 boxes a-year more than was produced by both fisheries while the stake-nets were in operation.

With such facts properly authenticated we are really unable to swallow the selfish assertions of those sea Salmonicides who try to persuade us that, if they be permitted to supply our tables, we shall never lack salmon. On the contrary, we have a shrewd guess that if their purses be replenished they will not carry their philanthropy beyond the fish-market of to-day; if *it* be supplied, their purpose is gained. Posterity may sigh for salmon—what is that to them whose business is to catch fish for the men of this evil generation who insist on having salmon?

But we humbly opine that salmon are intended to last as long as men, and would so last if men were not such dolts as to deliberately transgress the laws of all-wise Providence And as Governments are for the common weal, and not for the special behoof of fishing proprietors along the coast, they do well in not risking the future existence of salmon, by leaving them to the consideration of those bent on slaying them by methods so exterminating as stake and stage nets are proved to be.

Upon the whole, then, though we cannot be in such hot wrath as he is (very naturally after losing, we have been told, £5000 in law expenses), we think that the late Laird of Ardross has much reason on his side, when with vigorous indignation he protests against the invasion of the rights of upper-river proprietors, who are made "clocking-hens to hatch the fish that the folks below are to catch and eat." Salmon, he argues, are not *feræ naturæ*, and belong to no particular property. "Each river possesses its own variety of the species which *belongs* to itself exclusively, and which is forced by instinct to return to it, and to no other river. The natural right of the owner of such river to its salmon is, therefore, just the same as the right of the owner of a bee-hive to its bees, or a dovecot to its pigeons." Precisely; but the difficulty lies in identifying such floating property. We have heard of a litigation as to a swarm of bees, which were finally returned to their rightful owner because he proved that he had dusted them with flour, and that those which had gone to his neighbour were white with flour. Salmon bred in rivers have undoubtedly been often identified on their return from the sea, and must frequently be caught in the sea with marks upon them showing in what river they were bred. But unless that river belong to a single proprietor, it is impossible to say to whom they belong. Mr Mackenzie argues, that if there were no fishings except in rivers, each proprietor of a salmon river would get his own salmon, the produce of his own property, and none else, for none else would enter it. As a general statement, this is true; but it is too strongly put. Accidental circumstances often force the salmon out of their natural course; and not a season passes in which stray fish from the Tay are not found in the Forth, or in the South Esk, and identified, in consequence of bearing the Stormontfield mark. But while it is not always practicable to identify the aquatic *estate*, as Mr Mackenzie terms it, of a river proprietor, he turns the difficulty of connecting the salmon with its owner into an argument against a maritime proprietor having a right to the salmon passing his property.

"How can he *connect* himself with them? They do not even trespass on his lands, like bees or pigeons, so as to afford so much as a *pretext* upon which to engraft a claim to them. If they migrate to the ocean, the ocean does not belong to him. The rights of his lands end at the water-edge. He has not a particle of right one inch beyond that, any more than any other individual of the community; or to the salmon passing in the adjacent water, any more than he has to the birds which are flying in the air. To talk, therefore, of his natural right, *ex adverso* (as lawyers say) of his *lands*, to the salmon belonging to the rivers, passing by an element to which his lands can pretend no right, is a perfect absurdity. He might as well claim, *ex adverso* of his lands, a right to the ships which are sailing past them. If the owners of the lands have any right at all, it must be a *legal* right, opposed to the *natural* right of the owners of the rivers."

Of this alleged legal right, derived by charter from the Crown, our indignant author makes short work in this fashion; which, however horrifying to lawyers, must make them sincerely lament the demise of Mr Mackenzie, for so long as one of such a spirit was in the flesh law-pleas must have been rife:—

"After the Crown has made a grant of a salmon river—or, what is nearly the same, the salmon-fishings of a river—it can be no more in the *legal* power of the Crown to authorise conterminous proprietors of lands or others to intercept the salmon proceeding to the river, than to intercept the water on its progress to the mill. The principle is exactly the same in both cases. We defy any casuist in the Parliament House to make a distinction between them. It is, therefore, clear that all the grants of salmon-fishings to coast proprietors, which, as we have shown, can only act by interception of the fish in their way to the rivers, to the manifest loss and injury of the river-fisheries, being all made in direct violation of a great and acknowledged principle of the law, are all necessarily, radically, and fundamentally *illegal!*"

In charity we hope that this was written on the morning after the Laird of Ardross heard of the *illegal* (?) decision of the House of Lords in the case Mackenzie *v.* Houston.

The blood of the Gael is evidently up; and he would think nothing of tossing in a blanket the Lord Chancellor, wig and woolsack into the bargain. But as we happen not to be in a passion, neither smarting under the crushing weight of a lawyer's bill, we see that really something may be said in favour of a proprietor's right to catch salmon in the sea which bounds his property. If they do not belong to him naturally—if no law can give him a right to catch them—the sea, by this mode of reasoning, is the private fish-pond to be specially reserved for the use of river proprietors, that salmon may therein grow fat, and not a fin of them be harmed till they choose to return to their native rivers. If salmon are bred in rivers, it is certain that their growth is in the feeding-grounds of the ocean; and most people will conclude that, to catch them *there*, when they are in the finest condition, must be *pro bono publico*, even though it may somewhat interfere with the profits of river proprietors. If Mr Mackenzie had sent to market a fat bullock reared at Ardross, he would, we presume, have been wrathful exceedingly if a neighbouring laird had claimed the sole property in the animal because it had been calved on his land. If rivers supply salmon with nurseries, no sane man should jump to the conclusion that, *therefore*, they must never be caught in the sea, where alone they attain that adipose maturity which makes them worth the catching.

Our disappointed litigant treats of the "Rights of Parties," but in such hot blood that, to his disordered vision, there is but *one party*—the proprietor of a river-fishing. The really valuable part of his book is not that on rights of parties, but that on which he comments on Scottish statutes. We think it demonstrated that our old Scottish laws anent salmon-fisheries were infinitely more intelligible, more just, and more for the public advantage, than those decisions which regulate, or rather perplex, this department of our social economy. The courts of law nowadays seem to suppose that, unless stake-nets be planted in an estuary, or in the vicinity of a river, they can do little harm to the salmon. Moreover, though these fixed nets have been declared illegal in the Tay, this has not been followed by the removal of these destructive engines from all the estuaries in the kingdom. The Tay alone enjoys this happy exemption, which was gained by it being proved that *yairs* and similar engines are prohibited by the ancient statutes, "in waters where the tide ebbs and flows." It took fourteen years to determine whether "waters" meant waters in general, or only rivers! "The Judges," Mr Mackenzie informs us, "quoted old songs, and the counsel cited the classics, to show the intention of the Scotch legislators in framing their fishing-statutes, until at length, by their united efforts, they made *nonsense* of the statutes, and then most unjustly laid the blame, not upon themselves and their absurd constructions, but upon the defunct legislators." The grand debate was regarding these words, in the first statute of Robert I., from which all subsequent statutes appear to have been framed. *Yairs* and all fixed engines are prohibited "*in aquis ubi ascendit mare et se retrahit, et ubi salmunculi, vel smolti, vel fria alterius generis piscium, maris vel aquæ dulcis, ascendunt et descendunt;*" that is, in waters where the fry of any kind of fish is to be found. It certainly appears incomprehensible how the Court of Session should have held that this refers only to *rivers*, where yairs never are found, because they can only be set where the tide ebbs and flows; and that *waters* mean rivers, where, assuredly, the fry of *sea* fish cannot be found. The Lord Chancellor, in the Kintore case, remarked that the words "*ascendunt et descendunt*" appeared to denote the ascent and descent of the fry of salmon in *rivers.* But it so happens that salmon-fry *never ascend* a river. As soon would a feather go against the wind as salmon-fry ascend a river; instinct hurries them *down* to the sea whenever they reach the stage at which they are termed "*salmunculi vel smolti:*" and that the ancient Scottish statutes intended to prohibit fixed engines *in the sea* is proved by the statute 1488, which ordains "That all yairs and fish-dams that are in *salt* water, where the sea ebbs and flows, be utterly destroyed, as well those that pertain to our Sovereign Lord as all others throughout the realm; and that cruives set in *fresh* water be made according to law."

If the observance of this ancient law were universally enforced, if the capture of salmon at sea were permissible only by net and coble, there would be an end of the long feud between the salmon-slayers, inland and maritime; and perhaps, since the old tacksman of Ardross, after escaping the meshes of the law, has at last been caught in the net of the great destroyer, his son and successor may be so far appeased as to say to his father's whilome successful opponents, " *Pax vobiscum*—Weel may the boatie row." And yet we are not sure about the present Mr Mackenzie being amenable to reason. He has crotchets in his head, to dismiss which will not be easy. He believes that cuckoos, stone-chatters, and water-wagtails hybernate; and notwithstanding the very numerous modern instances demonstrating that smolts return, as grilses, in about fifty days after their arrival at the ocean, he also maintains that " the smolt requires one year at least to perform its migrations, to the North Sea it is supposed, and that at the expiry of this time it returns *as a salmon.*" In order to carry out this whimsical theory, he sneers at the Stormontfield experiments as "ill-devised"—yes, ill-devised in his eyes, because in those of other people they demonstrate that his theories are dreams. For instance: of the salmon taken in the Tay during two years, it is known that nearly a tenth had been artificially bred at Stormontfield—that is, instead of being at the north pole, as they should have been according to Mr Mackenzie's hypothesis, they had been not only marked, but caught and eaten by the salmon-loving lieges. In order still further to carry out his whim, he is pleased to maintain that salmon and grilses are of different species, and that they do not spawn together. Now the intelligent and experienced superintendent of the Tay fisheries informs us that he has frequently seen them spawning together, and that when the net is drawn over the ford, for the purpose of stocking the Stormontfield ponds, salmon and grilse are dragged ashore when in the act of spawning, and manipulated upon promiscuously, the impregnation of the ova being alike successful whether between two full-grown salmon or between a salmon and a grilse. Mr Mackenzie also makes the astounding assertion, that having tied wire round the tails of grilse, he caught them three years afterwards still grilse. We can only attribute this to his not knowing how to observe; and therefore, with all deference to Mr Mackenzie, we shall expect smolts to become grilses, and grilses salmon, as they have been doing since the world began.

THE SALMON RIVERS OF ENGLAND AND WALES.*

However much we may wish it were true, inexorable facts forbid faith in the theory of the vegetarians. Science and experience demonstrate that the physical organisation of human beings, especially in northern climates, can be maintained in highest vigour only when they habitually consume a certain portion of animal food. "Carnivorous animals," Liebig declares, "are generally stronger, hardier, and more courageous than the herbivorous, on which they prey. The same difference is visible betwixt nations which mostly live on plants, and those which live on flesh. No aliment acts so rapidly as flesh in the reproduction of flesh, and the reparation, at the cost of little organic force, of the muscular substance wasted by labour."

In this country the consumption of animal food by the mass of the people is no doubt much greater than in France, still it is far from sufficient; and as the difficulty of procuring it in adequate degree is constantly increasing, we know of no more important political problem than that which relates to the procuring of it in such proportions as are requisite for the main-

* 'Report of the Commissioners appointed to Inquire into the Salmon Fisheries' (England and Wales). Presented to both Houses of Parliament by command of Her Majesty: 1861.

tenance of the physical energies of the people. A half-starved nation is robbed of half its strength; and criminal statistics demonstrate that the public peace and morality have no greater enemy than hunger.

All this we have been urging for years past in communications on the physical condition of the people, and the means of public alimentation. We have been laughed at for our commendation of horse-flesh as a palatable and highly-nutritive article of human diet; but we rejoice to know we have not in vain pointed out our fisheries as an inexhaustible means of supplying the wants of our rapidly-augmenting population. With "Scotch Salmon and Scotch Law" we have made our readers so well acquainted that on that wide field we do not purpose again to enter, until some new legislative movement attracts public attention. We have heard that the Lord Advocate, discouraged by his recent failure to pass an Act regulating Scottish salmon-fisheries, is not disposed at present to trouble himself with that vexed question. But that there will be legislation as to the salmon rivers of England and Wales can hardly be doubted. It will be a scandal if no action follow upon the Report of the Commissioners, whose inquiries have revealed a state of matters demanding the instant consideration of the legislature, if one of our sources of national wealth is not to be wholly destroyed.

Matters are bad enough with our Scottish rivers, with the exceptions of the Tay and the rivers owned by the Duke of Richmond; but those of England and Wales are in a state of unproductiveness far more deplorable. Most of us in Scotland know so little about these rivers, even by name, much less as haunts of the salmon, that we are confident our readers will share our surprise at finding the Royal Commissioners reporting upon no less than a hundred and sixty-five. With the solitary exception of X, every letter of the alphabet is represented in the initial letter of some Welsh or English salmon river, once famous, but now shorn of its glory, and mourned over by the followers of the net and the angle.

The rivers of England and Wales exceed in extent those either of Ireland or Scotland, which yield large quantities of salmon, and rentals which to the river-proprietors of England appear enormous. They embrace a full average proportion of water suited for the propagation of fish, with rapid streams, deep pools, and good gravelly spawning-beds. "I have fished," says Captain Grant, "some of the finest salmon rivers in Europe and America, and, having killed many thousands with the fly, I may be supposed to know something of the habits and natural history of this noble fish. My experience convinces me that the river Taw might be made one of the best, if not the best, salmon rivers in England." Another witness depones: "I agree with Captain Grant that the Taw is one of the finest rivers in the kingdom. I almost believe that the Taw and the Torridge, if properly watched, would supply the metropolis with salmon." As to the former abundance of salmon in the Taff, we find a fisherman declaring that, seventy-four years ago, his father and two brothers took 500 salmon out of one pool: "and it was a common custom for me and my brothers to make a slip-knot in a piece of salmon twine, and put it round the tail of a salmon, tie the other end to a busbard, and let the fish swim about the pool till we wanted it."

The evidence of Mr Thomas Ashworth, a practical pisciculturist of great enterprise and experience, is not less decided as to the natural capability of English rivers for the production of salmon:—

"The Irish Fisheries Commissioners, in their Report for 1857, state that the value of the salmon caught in Ireland is above £300,000; whereas, in England, I have never seen any calculation that made the annual produce amount to £10,000 in value. We have in England and Wales about one-fourth part more rivers in extent than there is in Ireland, consequently it is reasonable to assume that the English rivers may become equally productive with similar laws and appliances. By calculation we find that the river Shannon contains an area of 7200 square miles, and produces salmon of the annual value of £80,000. The river at Waterford contains an area of 4000 square miles, and produces salmon of the annual value of £40,000. Then I will take two Scotch rivers. The river Spey produces an annual income of more than £12,000, and it contains an area of about 1050 square miles. The river Tay contains an area of 2800 square miles, and produces salmon which I estimate at the annual value of £30,000, the fishery-rents this year having increased to £14,000. The river Ouse and its tributaries contain an area of 2400 square miles, and probably produce salmon of an annual value of £400, the rent being £122, 15s. 0d."

When we know that the annual value of all the English salmon rivers is three times less than a single Scotch river, the Tay, we cannot but wonder. If we inquire how this has been brought about, the cause is obvious. As if it were impossible to exterminate the noble salmon, we learn from the Report of the Royal Commissioners that it is cruelly and senselessly persecuted in every stage of its too short life. Under the names of "last spring," "penk," "samlet," the salmon-fry, when about to descend to the sea, are slaughtered in myriads by anglers, who sell them at sixpence a-pound to hotel-keepers, who are thus enabled to respond favourably to the tourist's query, "Waiter, have you any samlets for breakfast?" At a small inn near St Asaph four men were accidentally seen seated at a table, with a tray before them, piled up eighteen inches with salmon-fry, which they had caught with a net, and which they were expertly gutting and trimming, as the first step in the process of conversion into "potted char" for the Liverpool market! A still more frightful destruction is indicated by credible witnesses declaring that they know instances of bushels of salmon-fry given to pigs, and employed as manure for the garden and the field. A witness, who was horrified at seeing a man manuring his garden with salmon-fry, coolly informs the Commissioners that he thinks a salmon good within a fortnight of spawning, and that on the Taw "you will hear many gourmands say, 'Where is the pea?' They look for the pea. I assure you that is the case here." At Bideford, it seems, the pregnant fish is reckoned good the day before she spawns! This depraved taste is the reason why the population on the banks of many English rivers show no mercy to the breeding fish. Declaring them "very good," a state of lawlessness in pursuit of them is sometimes witnessed in Wales, throwing into shade the sufficiently highhanded doings on Tweedside. "Rebecca and her daughters," with whose antipathy to turnpikes we are familiar, are remarkable for their unseasonable love of salmon. So recently as December 1857, a crowd of these desperadoes, who are mostly small farmers, and their sons and servants, assembled with blackened faces, and, armed with guns, proceeded to devastate the fords on the little river Elan, which they would have effectually accomplished but for the promptness of the river conservator, who, rushing to the fords, disturbed the fish a few minutes before the approach of the enemy.

It is deplorable to learn that, in certain parts of Wales, the law is powerless to repress such murderous poaching, "all the country being up in arms, so that you could hardly fix upon anybody;" and a gentleman, anxious for the preservation of the fish, dared not come to give evidence before the Commissioners;—"there is such a gang of people about, who make a regular trade of killing the salmon, that he will not come forward; he is afraid that his cattle will be killed, or his stacks burned."

Besides the fondness of the gourmands for "the pea of the salmon," the pregnant fish is killed for the sake of the value of her roe as a bait, which is sold at two shillings a-pound, and upwards.

That salmon should have decreased in England under the operation of such injurious influences is no marvel; and the misfortune is that the temptation to slay them out of season is on the increase. The Parisians have acquired a taste for British salmon; and whether it be owing to ignorance of our law regarding salmon out of season, or to the talents of the French cooks in disguising unseemliness, it is certain that tons of British salmon find their way to the French metropolis in such a state as ought to make them be deemed unfit for food. The price—about three francs per pound—is so remunerative as to tempt dealers in fish to send unseasonable salmon, "on the sly," in baskets with legs of hares sticking out, in order that the contents may pass for game; or lyingly labelled "Clothes—to be kept dry;" or, "Glass—this side to be kept up."

In fact, this nefarious traffic is so organised as to have a trade nomenclature of its own. O.S.S. stands for old salmon stuff. As there is now a law prohibiting the exportation of fish out of season, this traffic can be put down; but not easily, as it is carried on secretly in order not to come under the notice of the inspectors at Billingsgate, from whose interference it has little to apprehend, as their instructions restrict them to seizing food "not wholesome," which with them is not synonymous with out of season; so that one of them declares that, however large with roe a salmon might be, he would not interfere if it had a healthy appearance.

This indiscriminate slaughter of salmon, young and old, has been going on in England for nearly six hundred years at least, and in defiance of the law. The first Act of Parliament for their protection was, we believe, the 13th of Edward I., 1285. It relates to the Ouse, the Humber, and other rivers, and enacts that "all other waters wherein salmons be taken shall be in defence for taking of salmons from the nativity of Our Lady unto Saint Martin's Day;" and the destruction of "young salmons" is forbidden "from the midst of April unto the Nativity of Saint John the Baptist"—that is, the 24th June. That the salmon should have the worst of it in this long contest with human greed and folly is quite intelligible, even if this were all that we know of the adverse influences under which they have contrived to perpetuate their species. But who shall recount the perils which beset the life of a salmon? Every river almost can boast of having within or along its course some means of destruction peculiar to itself for ensnaring the luckless salmon. The estuary of the Severn has its fixed engines — known as "putts," "putchers," and "trumpets." The rivers running into the northern side of the Bristol Channel have "coracles" and "stop nets." The estuary of the Ribble is swept by the "wing net," and that of the Parret by "butts."

In the fresh water the salmon has to encounter a new range of fixed engines, designated as "cages," "coops," "slaughters," complained of from the days of old, but more or less actively destructive in all English rivers to this day, notwithstanding sundry Acts of Parliament directed against those owners of salmon-fisheries, who, as one of Queen Anne's Acts declares, "regarding only their private and greedy profit, do destroy the stock of the said fisheries by preventing the breed of good fish to pass in season through their fishing wires and fishing hatchways from the sea into the rivers to spawn, whereby not only the increase of the said fish, but also the growth thereof, is in great measure destroyed."

These causes of destruction to salmon originate in *malice prepense.* There are many, besides, springing up with advancing civilisation, and not designedly, yet ruinously injurious to fishes. For instance, the "silver Thames" of our oldest English poets is now the stinking tidal sewer whose odours penetrate into the Houses

H

of Parliament, and compel our legislators to sympathise with the unfortunate salmon, which, until 1824, strove to perpetuate their race in these polluted waters. Of no degenerate stock were the Thames salmon. In the journal of Richard Lovegrove, tenant of Boulter's Lock on the Thames, we read:—" 1801. This year there were taken 66 salmon, weighed 1124 lb. Two salmon weighed 65 lb., and three that weighed 70 lb." As the custom of sending fish to London in ice had not begun, the Thames salmon of those days used to be sold at 10s. or 12s. a-lb. Is there a probability that the Thames may be restocked with salmon? We think not. So thoroughly is it polluted by the filth, the gas-works, and the steamers, that we are persuaded that no salmon, however impelled by philoprogenitiveness, could push its way to the spawning-beds in the upper parts of the river. We have it in evidence that salmon have been seen to turn tail the moment they came within the influence of a certain mining abomination in a river in Wales. And having had no small difficulty in resisting the nausea occasioned by sailing down the Thames, we cannot conceive that the dainty salmon should either attempt or survive swimming up the odorous stream. As to the tender smolt effecting a passage to the sea through such pestilent waters, that appears equally impossible. Mr Gould, the naturalist, is of a different opinion, and recommends having recourse to artificial propagation. "If," says he, "10,000 smolts went down to the sea, possibly 100 grilse, or some small proportion, would return;" and he adds that Mr Noble, of Maidenhead, is ready with 200 guineas to assist in bringing back salmon to the Thames. Did he ever reflect whether they would be eatable, supposing them reintroduced into such a river! The Commissioners' Report brings out the curious fact that the pollutions of rivers, when they do not destroy the fish, yet communicate to them such a detestable flavour as to render them uneatable.

The Mayor of Gloucester, in answer to the question,

"Are you aware of any noxious matter being discharged into the Severn?" replies, "There was some refuse from creosote-works discharged into the river, and it was supposed by every one that it had a very injurious effect on the Severn fisheries. I know that the fish tasted very much of the creosote when we did get one for dinner." As to the effect on fish of gas-refuse, another witness depones that "they tasted just the same as the tank timber smells." Tar appears to have a very special power of tainting fish without injuring it externally, so that the epicure only knows when it is cooked what a precious mess is a tarred salmon. This is no fancy of the Mayor of Gloucester. A fishmonger there, having a salmon returned by a customer, is honest enough to confess, "I tasted the tar itself throughout the fish from head to tail. There was no mistake about it; it was like tar itself."

This should discourage the attempt to reintroduce the salmon into the Thames. If successfully made, the result will be the capture of fish having a composite *goût* corresponding to the number of the savoury ingredients in the midst of which it has lived. And if the Mayor of Gloucester be invited to an aldermanic banquet in London to celebrate the reintroduction of salmon into the Thames, we predict that he will not envy Messrs Gould and Noble the dearly-purchased *bonne bouche*.

We therefore hope that discredit will not be thrown upon the really valuable system of artificial breeding by the attempt to make it the means of restoring salmon to the Thames. Let this river be made to produce those fish for which it is still found suitable. Under the care of an angling association the breed of fine large trout has increased astonishingly. As bait for large turbot and the immense Dogger Bank cod, there is a great demand by Dutch fishermen for lampreys. "Sixty thousand went the other day, at £3 a thousand, from Teddington." Well, it is some consolation to learn that, in any sense, we have still a "silver Thames." We wish our Scotch fishermen would learn the value of eels. Those who wish to know it will find interesting information in "Maritime Pisciculture."

The cost of cleansing the Thames, so as to again render it habitable by salmon, would be so enormous that we despair of it being deodorised. This dismal result of forgetting Lord Palmerston's definition of dirt —"valuable matter in an improper place"—is a warning to the dwellers on all rivers to jealously preserve their purity by insisting that all towns and villages shall find other means of disposing of their filth than by casting it into running streams. If the Lord Advocate cannot carry through Parliament a bill regulating Scottish salmon rivers, it is to be hoped that the legislature will not refuse to take measures to hinder them being defiled, and so becoming nuisances like the Thames.

The salmon has been banished from several rivers in Wales and Cornwall by poisonous matter from mines— particularly lead-mines. A total extinction of animal life has taken place in the waters of the Rheidol and the Ystwith; and even the sea-fishery, to the extent of several miles from the influx of these rivers, has been much deteriorated from the same cause. It has also been proved that not only the fish in the rivers, but animals grazing on their banks—cows, horses, pigs, and poultry—had been poisoned, not so much by drinking the water as by eating the grass, which, during floods, had been covered by the infected waters. Many acres of land have thus been rendered worthless. The mischief occasioned by water from lead-mines sometimes leads to unexpected results. The birds in an aviary perished unaccountably, until it was suggested that they were supplied with sand impregnated with lead. The sand being changed, the mortality ceased. In Cornwall, which is peculiarly a mining county, the salmon-fisheries may be said to have been virtually destroyed by the mines. Lead-works are about to be worked on the Wye, to the great alarm of the proprietors of salmon-fisheries; and the same apprehension exists as to the Conway, and other rivers of great natural capabilities. We regret to perceive that the Commissioners deem the extinction of the Cornwall fisheries inevitable. They are fortunately not of the highest value, and to prefer them to the great mining interests which form the staple industry of a wealthy county, would no doubt be to preserve salmon at a preposterous cost. Still the question recurs, Is it right that, without compensation, mines should be permitted to poison rivers? Here there is evidently much troubled water in which the lawyer will know how to profitably ply his craft. And if there be any means of rendering innocuous these poisonous effluxes from mines, we hold that the legislature should insist upon their being employed. We are glad to find M. Moggridge, Esq., expressing a confident opinion that the construction of filters might not only save the destruction of animal life, but prove remunerative to the miner, by enabling him to save mineral ingredients which he now throws away. On the Towy he maintains that things go on as they did in the quicksilver mines of India, where, for centuries, they threw away the most valuable part of their produce. The comparative amount of capital employed in manufactures renders it difficult to hinder the pollution of salmon rivers without hampering trade and industry in a manner which the Commissioners deem unjustifiable for the sake of preserving salmon. " The interests of manufactures, nationally considered, must be deemed paramount to those of fisheries; but in like manner, as in the other case, we consider that a watchful eye should be kept over the introduction of new causes of pollution; that the rivers hitherto uncontaminated should be kept from harm; that all such nuisances as may without undue sacrifice be preventible, should be prohibited, and, if they arise, checked in the outset." On these principles salmon are doomed to extinction in many at present valuable rivers; and being unable to resist the force of the utilitarian argument, we can only groan under the prospective loss. We are slightly comforted by learning that most of the Acts under which gas companies are constituted provide penalties against suffering any escape from the works into rivers; and that, as the residuary matters are valuable, the Commissioners recommend that the law against this abuse be made peremptory.

The recommendations of the Commissioners in providing a remedy for the evils they deplore are mainly those indicated by the Select Committee of the House of Lords. Considering the importance of an increased supply of valuable food for the public, they recommend the appointment of a body of commissioners, charged with the conservation of rivers; and regarding the salmon fisheries as national property of great value, they are persuaded that a moderate annual cost incurred for their reclamation would be amply compensated to the State. Though no allusion is made to the action of the French Government in this department, we point to it as affording ample justification of Government interference. In our articles on Pisciculture we have narrated how the depopulated rivers of France have been made to teem with valuable species of fish, and how the ruined oyster-beds along the shores of France have been restored by a very moderate expenditure of the public money. So very much may be said in favour of the appointment of a Minister of Rural Industry, that we should be glad, in this respect, to see our Government imitating that of France. It is curious that the Indians inhabiting a district of British Columbia, a region abounding with salmon, should have an institution such as the French have, and such as we desiderate for this country.

Mr Kane states that these people are governed by two chiefs—the Chief of the Earth, and the Chief of the Waters. The latter personage appears to be of great importance. No one is allowed to fish without his permission. He distributes the fish taken during the season amongst his people; every one, even the smallest child, getting an equal share, which is not stinted, seeing that the chief informed Mr Kane that in one day he had taken in his basket as many as 1700 salmon, weighing on an average 30 lb. each, and that the daily average capture was 400.

Our Chief of the Waters, when we get him, would not, it is true, be so paternal in his distribution of the good things pertaining to his domain, yet a most useful public functionary would he be, were he to exercise a little salutary despotism over our fisheries. One part of his duty might be to order the removal of all obstructions to the salmon reaching the spawning-beds; or, at least, the providing them with facilities for doing so by means of "ladders," up which, it is proved, salmon will ascend as regularly as domestic fowls clamber to their roost. The cost of this useful invention is trifling. An ascent of the unusual height of fifteen feet cut through the rock, moreover, can be provided for £60. As there is no loss of water-power, it is most reasonable that every weir or obstruction of any kind should be demolished, if the proprietor will not provide a ladder for salmon. The Commissioners are so considerate as to propose that he shall do so at his own expense only if his weir be of recent erection. They are willing that this appliance shall be furnished to ancient weirs at the cost of the local fishing-boards. As the introduction of ladders into the rivers of Ireland has been effected easily, and with the greatest advantage, we hope proprietors of Scottish salmon rivers will not wait for the compulsitor of an Act of Parliament, but will at once introduce this ingenious contrivance of the late Mr Smith of Deanston.

Perhaps the most important recommendation of the Commissioners relates to the fixing of a uniform close-time. At present this may not exceed 150 days; but the law fixes no minimum time, and the dates of opening and closing are fixed by the quarter-sessions, with a varying latitude productive of the greatest anomalies and the most serious injury.

The great breeding season in England and Wales of all fish of the salmon kind is in the months of November, December, and January. The same season extends to the rivers of Ireland and Scotland—the great rush of the fish to their breeding-waters taking place in the end of September and the beginning of October, and increasing with the advancing season. Diversities in the spawning-time are naturally within so limited a time that this time may be considered as nearly uniform throughout the United Kingdom. The so-called "lateness" of certain rivers is greatly caused by the period during which they are fished; and the characteristic of earliness may be lost by any river which has not the benefit of an early close-time. If the early-spawning fish be captured, those permitting the abuse have themselves to thank for the compelled acknowledgment, "Our river has become later than it used to be." This has been proved in Ireland, where the early closing of rivers has produced a corresponding early supply of salmon in certain rivers. The common fishermen in Ireland were at first much opposed to early closing, but now even they acknowledge that the fishing proprietors are right, when, in some instances, they have petitioned the Commissioners to exercise their power of closing the rivers at even an earlier period. Our rivers can never prosper till closed at an early season; and poaching and the sale of unseasonable fish cannot be repressed until the possession or sale of salmon after a certain fixed date is declared illegal.

As to fixed engines on the estuaries and the seacoast, the Commissioners recommend their total suppression, for most cogent reasons; thus going beyond the recent bill for regulating the Scottish fisheries, which only proposed the removal of stake-nets to a certain distance from the mouths of rivers.

We trust that their valuable Report will receive the attention which it merits, both from the legislature and the public. It is a mistake to suppose that this is a matter chiefly interesting to the rich, or to the proprietors of fishings. While the population is rapidly increasing, and every effort to keep up the supply of food is barely sufficient, it is surely most unwise to neglect or destroy the produce of our rivers, which may easily be made to yield a greatly increased revenue, as well as much palatable and nutritious food at a price within the reach of the great body of the people. Only eighteen months ago the working-classes in certain English towns came to a resolution to purchase no butcher-meat until its price was lower. To multiply salmon so that it shall be reduced to the ordinary price of meat—to stock our rivers with various other fish of considerable value—to induce the formation of fish-ponds, such as were common in this country before the Reformation—to accustom our people to the more liberal use of fish as a common article of diet,—ought to be the aim of all who know how important it is that the physical energies of the nation shall be maintained by the mass of the people being adequately fed. How to supply our rapidly-increasing population with butcher-meat is a problem not easily solved. London, with its all-devouring maw, swallows every year about 270,000 oxen, 1,500,000 sheep, and 30,000 swine. The total value of all the flesh, dead and alive, imported into London, cannot, according to a recent estimate in the 'Scottish Farmer,' be much less than £14,000,000 annually. It is evidently high time that the people in the country should in fish find a substitute for flesh, seeing that this monster city swallows so much of "the roast-beef of old England," and is not satisfied without levying heavy weekly supplies of oxen from the Land of Cakes.

SALMON—BRITISH AND COLONIAL.*

IN "Scotch Salmon and Scotch Law" we pointed out the important interests affected by the decision of the House of Lords in the case of Gammell v. the Commissioners of Woods and Forests. Having explained that the Crown was declared proprietor of all salmon-fishings on the coast of Scotland, except so far as these had been granted to the Crown vassals by charter, we proceeded to observe, "It is reasonably inferred that the Crown will not permit this noble heritage to lie waste, but that it will stimulate salmon-fishing on the coast, by granting charters for a consideration, the amount of which may largely swell the public revenue. Supposing this course to be followed, the question of stake-

---

* 15th March 1861 : 24th Vict. A Bill for the Improvement, Regulation, and Protection of the Salmon-Fisheries in Scotland, and on the Coasts of the Solway Firth, and in the Mouth or Entrance of the River Tweed.

The same Bill (as amended by the Select Committee), ordered by the House of Commons to be printed, 5th July 1861.

'Report of the Commissioners appointed to Inquire into Salmon-Fisheries (England and Wales), together with the Minutes of Evidence.' 1861.

'Reports by the Select Committee of the House of Lords upon the Salmon-Fisheries of Scotland.' 1860.

'Salmon-Fishing in Canada.' By a Resident. Edited by Col. Sir Jas. Edward Alexander. With Illustrations. London, 1861.

net fishing must be considered in all its varied bearings upon the rights of individuals and upon the interests of the public."

How truly we foretold the result of this important decision, is already apparent in the Salmon-Fisheries Bills for England and Scotland, and in the Reports by Select Committees both of the Lords and Commons. By the publication of the evidence taken before these committees we are made acquainted with the view which Government takes of its responsibilities to the Crown and the public after the decision in the case of Gammell. Before going further, it may be for the benefit of some of our readers that we indicate the facts in this case, as well as explain the legal status of a Scotch salmon.

Mr Gammell granted a lease of the salmon-fishings *ex adverso* of his estate of Portlethen, Kincardineshire, to Messrs Gray and Hutcheson of Aberdeen and Findon, who had been in the habit of using stake-nets and other engines for catching the fish. The Commissioners of Woods and Forests intimated to them that this was illegal, and offered them a lease at a moderate rent. The offer being declined, they raised, in 1849, an action against Mr Gammell and his lessees, claiming to have it found—1st, That all salmon-fishings round the coast of Scotland formed part of the hereditary revenues of the Crown, except so far as they had been granted out to the Crown vassals by charter; 2d, That as Mr Gammell had no such charter, the lessees had no right to fish for salmon on the coast adjoining his property. The Court of Session, with the exception of Lord Justice-Clerk Hope, found and declared that all the fishings round the coast of Scotland belonged to the Crown, as part of its hereditary revenues. An appeal to the House of Lords confirmed this finding. When giving judgment Lord Campbell observed (and we pray that those maintaining the legality of stake-nets would mark the statement) :—

"The interdict granted by the Court below struck at all kinds of apparatus for which the land was required as a point of support, and to that extent he agreed with the Court, and held that, so far as stake-nets and other apparatus had any connection with the land or sea-coast, these were illegal. In short, the kind of fishery of salmon which was prohibited was that to which the use of the sea-coast was essential. The right of the Crown in the present case must be rested entirely on the authorities peculiar to the law of Scotland, and it was needless to refer to the law of England, or of any foreign country, for the purpose of showing how such rights were viewed there. It seemed that early in the history of Scotland salmon was singled out as distinguished from other fishes, and the right to fish for it was classed *inter regalia*. So Craig, Stair, Erskine, and Bell laid it down that, unless there was a special grant, or a general grant followed by possession, the subject had no right to fish for salmon."

Further light on the legal status of Scotch salmon is afforded in the Lord Advocate's evidence before the Lords' Committee.

" In Scotland the salmon is not a royal fish in the proper sense of the words. It is not like sturgeons and whales, the property in which truly belongs to the Crown. The whale was held to be a royal fish, but a salmon-fishing is a separate estate. It is a separate, real, or heritable estate, as it is called in the law of Scotland, and consequently is vested, as all real estate is, in the Crown. By the law of Scotland, under the old feudal principle which still subsists, all the land and real rights of Scotland are vested in the Crown, and can only pass from the Crown by a specific grant, or by what is equivalent to a specific grant. The right of salmon-fishing, accordingly, is a separate estate—so much so, that the Crown may grant a right of salmon-fishing in a river although the grantee have no property on either bank. Therefore salmon-fishing can only be exercised by a person having a right from the Crown, but that need not necessarily be a direct grant of salmon-fishing. It is sufficient if there is a title from the Crown of fishings generally, *cum piscationibus* or *cum piscariis*, followed by possession for a prescriptive period, which is forty years, or time immemorial. But rod-fishing, or mere angling for salmon, is not sufficient to make a prescriptive title in the absence of a direct grant; but it is necessary that the fishing should be exercised by some mode of mechanism, like cruives and yairs, as they are called, or nets; that is to say, fishing with net and coble, or fishing with cruives and yairs, following on a general grant of fishings for the period of prescription, will be sufficient to constitute a prescriptive title, but not rod-fishing or spearing."

Salmon-fishing having thus undeniably been feudalised, Mr Gammell had manifestly no right to fish for salmon, because his charter only mentioned white fishing, and by the ordinary rule of construction excluded every other. It being thus established that the Crown's right to salmon-fishing in the sea extends to three miles from the coast, and is not merely a trust for the public, but a patrimonial right which the Crown may use and make money of, the Crown had to determine what use should be made of so valuable a right. Most fortunately, as we think, for the public good, the decision arrived at was to exercise it entirely for the benefit of the salmon-fishings.

"I do not think it right," observes the Lord Advocate, " that the Crown should attempt to make a revenue from the salmon-fishings, excepting in the way in which it can be done most favourably for the preservation of the breed of salmon. It is a very important matter to the salmon-fishings that the decision was given in that way, because it puts it in the power of the Crown to do a very great service to these fishings. If it should appear that bag-nets in the sea are really destructive to salmon-fishings in the rivers, which is a very general opinion, I should think that the Crown should prohibit, in all leases to be granted of salmon-fishings in the sea, the use of these fixed engines. The best use which the Crown could make of the rights would be to grant leases to those interested in the river-fishings, with authority to fish by the net and coble, and with a prohibition against fishing by fixed engines, and an obligation to prevent others doing so. I think if the Crown rights were exercised in that way it would be most beneficial to the salmon-fishings. On the other hand, I should be sorry to see a system of granting leases to private proprietors of fishings in the sea, for inconsiderable rents, so that the person taking the leases would have no interest in the river-fishings at all, and only be desirous to catch as many fish as he could."

These are the views to be expected from an impartial and enlightened legislator and a high public officer well acquainted with the value of the Scottish salmon-fisheries as an important branch of national industry, crippled, unfortunately, by varying legislation and the thoughtless cupidity of those killing salmon at all seasons and by all methods. We have for several years been giving utterance to views very similar, and quite as disinterested, seeing that our connection with fisheries is simply that originating in a predilection for this department of natural history. We are gratified to learn that these are not merely the private opinions of the Lord Advocate, but also the views of the administrators of the property of the Crown, who have resolved that its rights shall be exercised only for the public good. And considering the history of those fishery proprietors whose pecuniary interests have been affected by recent inquiries, we think it well that the public should know that attention to its interests was the origin of these inquiries. The proprietors along the Scottish coasts have hitherto observed no close time, and, we understand, have contributed nothing for the protection or improvement of salmon-fisheries, but have maintained that it is both right and according to law that they shall be permitted to capture salmon in the open sea in any way they choose. The Lord Advocate is assuredly right in viewing the establishment of the Crown's property in sea-fishings as a favourable opportunity for settling the modes in which these should be conducted. As no less than 600 such fishings exist on the coast of Scotland, the Commissioners of Her Majesty's Woods and Forests, whose inquiry ascertained this fact, could not avoid a further investigation for the purpose of ascertaining whether it is just and expedient, having regard to the rights of the Crown and individuals, that the existing mode of fishing by bag and stake nets should either be done away with altogether, or should be regulated. And yet, for their most reasonable desire to discharge their duty to the Crown and the public, the Commissioners have been represented as rapacious interferers with the vested rights of the said 600 *riparian* proprietors, as they are termed, who insist that their stake-nets shall not be meddled with, at least without compensation, seeing, as they assert, that their right to use them, however doubtful at one time, has now the validity derived from prescription.

In order to show the reasonable spirit of the Commissioners, we quote the following, from the evidence before the Lords of Horace Watson, Esq., Solicitor to the Commissioners of Woods and Forests and Land Revenues :—

" To say that bag and stake nets should be prohibited altogether in Crown leases would be deciding the very question which the Committee are going to investigate —namely, Whether they should be done away with at all, or whether they should be regulated or abolished in certain localities? I may add that the Commissioners of Woods are bound, by their duty to the Crown and to the public, not to give away the Crown property. It is their duty to see that a fair revenue is obtained from it. They might no doubt have derived a larger revenue from the fishings already, if it had been intended to view them simply as a source of revenue; but, with the express object of preventing that being done, the course taken has been to grant leases, since the decision in Gammell's case, only for a term of three years, subject to express notice that the grantees should take subject to any alterations that Parliament should make in the mean time."

Unless the stake-net proprietors advocate the absurd theory that six hundred distinct fisheries for salmon shall be conducted as they please, irrespective of all questions concerning the habits of the fish and the interests of the public, they should not have objected to investigation in the reasonable spirit indicated by Mr Watson. And when the result of investigation before committees both of Lords and Commons, as well as before a Royal Commission, was the production of a conviction that it was the duty of Government to introduce a bill abolishing all fixed engines for the capture of salmon, we certainly did not anticipate that the design of the Government should be frustrated, and that the bill should turn out to be a legal sanction of modes of fishing recently introduced, and only tolerated with many misgivings as to their expediency, and even as to their lawfulness, according to the spirit, if not the letter, of the ancient Scottish statutes.

Such a result has been avoided, in the last stage of the bill, by its despairing opponents having the good luck to be able to shelve it for the session, by a "countout" of the House of Commons.

As we have not seen the last of this bill, we, the salmon eating and catching public, as well as the proprietors of salmon-fisheries, along shore and inland, have such interest in it that it is worth our while to acquaint ourselves with its provisions. Of many of these we cordially approve, but there are several which we hope to see excluded.

The bill, as amended by the Select Committee, differs in most important respects from that brought in by the Lord Advocate. At first the preamble ran thus :—
"Whereas it is expedient that the Acts relating to the Salmon-Fisheries in Scotland, and on the coasts or shores of the Solway Firth, should be consolidated and amended, and that further provision be made for the *abolition of cruives and fixed nets, and engines in rivers and in the sea*, the regulation of fisheries, the removal of obstructions, and the prevention of illegal fishing : be it enacted," &c.

The amended preamble proposes that "further provision be made *in regard to* cruives, and fixed nets, and engines in rivers and *estuaries*, and in the sea." The introduction of the word *estuaries* is a manifest improvement; but the omission of the word *abolition* shows that Government was defeated, and that the stake-net proprietors triumphed, in defiance of the recommendation of Select Committees of both Houses of Parliament, and of the Royal Commission for inquiring into the Salmon-Fisheries of England and Wales. As the withdrawal of the bill affords the opportunity of calmly considering its provisions, we shall endeavour to give an impartial *résumé* of the *pros* and *cons* in regard to those which admit of debate.

We must introduce our readers into the troubled waters of the contested portions of the Lord Advocate's bill, the foremost, of course, being that regarding stake-nets. What can be said in their favour? Mr Hector, St Cyrus, Montrose, and tenant of salmon-fishings in Scotland and Ireland, is strong upon the superiority of sea-caught salmon over those caught in rivers.[*] "For many years I had the largest retail trade in salmon in the city of Aberdeen. My supply was, for some time, chiefly from the river Dee; but, from repeated complaints of the softness of river fish, I had to bring sea-salmon from Montrose Bay to Aberdeen, to meet the

[*] 'Remarks on the Salmon-Fishings of Scotland.'

wants of the public, or to relinquish my retail trade. I had, therefore, to send sea-caught salmon a distance of thirty-two miles. I had to send the Aberdeen river-caught fish to London in ice. Preserved-provision curers will not take river fish if they can get sea fish to preserve." In corroboration of this statement, Mr Hector annexes letters from two salmon-boilers of long experience, which we should more highly value if they had not so much the look of being made up according to order. Though the one be dated "County Mayo, Ireland," and the other "Aberdeen," they are almost precisely in the same words, and these so peculiar as to generate the suspicion that both have the same origin.

We can easily believe that salmon from the upper waters of the Dee, if not despatched immediately, and not packed in ice, would not be so palatable to Aberdeen epicures as fish caught in the sea; and we are aware that salmon which have been several weeks in a river sensibly deteriorate. But again and again we have caught and eaten salmon on the Tay, thirty miles above the estuary, of the highest excellence; and, from personal knowledge, we aver that salmon travel this distance in the time betwixt Saturday night and Monday morning. In Mr Mackenzie's 'View of the Salmon-Fishery of Scotland' we find the following remarks on the idea that salmon taken in stake-nets are of a better quality than those taken in rivers:—"They might be better than what used to be formerly taken in the *higher* parts of rivers after they had remained long in them; but they cannot possibly be better than what is caught in the *tide-way* of the rivers, where nine-tenths of the fish are now taken. We would defy the most fastidious palate to discover a difference between them, or to distinguish the one from the other on the table. Mr Little, who was a great stake-net fisher, does not consider that there is any difference between a salmon taken in the sea or in a river, provided he is taken soon after he enters the fresh water. If he is taken in the course of a week after he enters the river, I do not consider him any worse; but few salmon are allowed to remain a week in the river during the fishing season." In a note, Mr Mackenzie adds :—" Salmon are always better for being a few days in their native water. It increases, like crimping, the firmness of the fish, insomuch that while a salmon caught in the morning in the sea is soft enough to be boiled and pickled the same evening, one caught in the fresh water retains its firmness, and would break in the kettle if boiled before next morning. The fish-curers or boilers, who are great epicures, always, accordingly, prefer for their own palates fish that have been some days in fresh water."

It is thus apparent that the quality of salmon taken at sea by the stake-net is not universally admitted to be superior to that of salmon netted in the river.

This question of quality may, in fact, turn out to be decided in favour of the river-caught salmon, if we consider the season during which the stake-nets are most successful—namely, during the months of June and July generally, but sometimes during May. The salmon from the month of May till spawning-time are gradually growing worse in condition, so that stake-nets cannot be said to capture the finest fish. Those taken abundantly in rivers during the months of winter and spring, being further from spawning, are necessarily in the finest condition. So that, *taking into consideration the whole fishing season*, even an epicure might give his verdict in favour of the superiority of river-caught salmon.

But even admitting the superiority of sea-captured salmon, we have still to debate the point whether they should be permitted to be taken by stake-nets.

The proprietors of these engines make their murderous success a reason why they should be permitted to exist. But we must remind them that it has never been the policy of our salmon-fishing laws to encourage salmonicide by the most destructive methods. Great anxiety has always been shown that this should be effected in modes compatible with liberty to the fish to fulfil its generative function in the rivers; and if it can be shown that there is ground for the general opinion that stake-nets impede the approach of salmon to their natal waters, where they seek to propagate their species, the Legislature ought to care for the common weal, and forbid the use of such nets, even though such prohibition shall injure the pecuniary interests of those who have asserted to themselves the right to use engines lucrative to them, but hurtful to other fishing proprietors.

Stake-nets and bag-nets being the fishing engines most objected to, we may inform some of our uninitiated readers that they are both constructed on the same principle; the only difference between them being that the one is fixed by stakes and the other by anchors and floated by corks. Stake-nets are always placed on sands, or ground left dry by the receding tide; but as bag-nets may be set in the deep water, and their leaders extended so as to embrace a whole firth, they are so much the more injurious. So many bag-nets have been placed in the sea near the mouth of the river Don, Aberdeen, that nearly three-fourths of the salmon are, it is said, intercepted before they reach that river, to the manifest injury of the river proprietors.

It is contended that the public is thus supplied with large quantities of salmon which would not be caught in the rivers, and that therefore the interest of the public demands the use of fixed engines in the sea. Every salmon caught in such engines has within it milt or roe, for the deposition of which, when the proper time arrives, it is impelled to make for some river, where it may be caught with all the certainty that can be desired by those not desiring the extermination of its species.

The more salmon caught at sea, the fewer can there

be left to carry on the generative process in the only waters where it can be pursued; and therefore an unlimited right to kill salmon at sea should not be permitted. To comprehend the reasonableness of the outcry against the exercise of this right by means of fixed engines, it must be borne in mind that until lately, and this only in particular localities, such as the mouth of the Tweed, stake and bag nets observed no close time, and were thus permanent obstructions to the riverward progress of the salmon. And when a mass of unimpeachable evidence has demonstrated their injurious effects on river-fishings, the impartial public must conclude that such modes of fishing should either be abolished, as was at first proposed by the Lord Advocate's bill, or carefully regulated, as the amended bill now proposes. Legislation there must be, because, notwithstanding the long and expensive litigations terminating in decisions by the House of Lords, the practical result is this, that salmon-fishing at sea by any mode, however destructive to the interests of the salmon-fisheries, cannot be put down by the existing law. The Lord Advocate having stated, before the Lords' Committee, that "it is the law that any mode of fishing is legal in the sea," the proprietors of river-fishings were wise not to accept the challenge to put down stake-nets, if they could, by an action at law. Their remedy is to be sought for in legislation preceded by inquiry; and it will not be sought in vain. Already their interests are essentially promoted by the kind of regulation which, it is confessed, stake-nets require. The provisions of the bill in regard to them, and all fixed engines for the taking of salmon at sea, are undoubtedly in the right direction. Whether, when the discussion before Parliament is renewed, the decision shall be in favour of "regulation" or "abolition," is somewhat dubious. But if the opinions of competent and impartial inquirers be allowed due weight, the decision, after all, may be in favour of abolition. We have not the slightest personal interest in the decision; and having exercised our unbiassed judgment, we cannot dissent from the Royal Commissioners (Sir William Jardine, Mr Ffennell, and Mr Rickards) when thus reporting to her Majesty:—

"We are prepared, after a full consideration of the case, to recommend the total suppression, by law, of all fixed engines on the estuaries and sea-coasts. The grounds upon which we have arrived at this conclusion are the following :—

" 1. These engines, with few exceptions, are of modern invention, and they are opposed to the whole aim and spirit of the fishing laws, the object of which was to secure the salmon a free passage to and from the sea, and to cause an equitable distribution of them throughout the rivers.

" 2. These engines are baneful to the fisheries, not only on account of the number of fish which they destroy, but also because they scare and drive them away to sea, when they come in shoals seeking the rivers, thereby exposing them to be injured or destroyed in a variety of ways."

While many are of opinion that the amended clause in the Lord Advocate's bill is a reasonable compromise between competing interests, the perusal of these deliberate statements of the Royal Commission will lead not a few to fear that this settlement of the vexed question has been arrived at from considerations not of strict justice, or of due attention to the natural history of the salmon. This fear is deepened on finding that the Fishery Commissioners for Ireland report that in 1860 the salmon-fisheries were not so productive as in the preceding year.

"The number of fixed engines in the tideways and on the sea-coast has increased, within seven years, from 270 to 386. This mode of capture has now extended to an abuse; but as it has been legalised by the Legislature, all that the Commissioners can do is to adopt as short an open season as the circumstances of each district or river require, and to enforce a strict observance of the close season."

Mr Hector, in his pamphlet already referred to, asks—
"Are not the proprietors of the shores of Scotland entitled to ask the British Parliament, not only to be allowed to have fixed nets in the sea, but to confer on them a title to the shores, such as has been readily awarded to Ireland?"

We answer, No. If the official conservators of Irish fisheries declare that the modes of capturing salmon on the Irish coast have extended "to an abuse," let us not clamour for the extension to Scotland of a bad law. If the Irish fancy that "justice to Ireland" requires that all modes of capturing salmon shall be lawful at sea, let us be more reasonable, and not so foolish as to hurt ourselves by importing a bad law from Ireland. The permission to use stake-nets, as stated by Mr Ffennell before the Lords Committee, was a compromise.

"We took the Act of 1842 sooner than not get any Act at all; but my opinion at the time was, that, for the benefit of the country, it should have been abandoned altogether, rather than sanction stake-weirs and bag-nets so far as they are sanctioned."

Though we gravely doubt the propriety of the compromise in regard to Scottish salmon-fisheries, we heartily approve of the determination to put down, by increased severity of penalty, all illegal modes of slaying salmon by the leister, or burning the water. It will be no easy matter, for many a day, we fear, to carry out the provisions of the law in these respects. Certain districts, especially in the south of Scotland, seem to be thoroughly demoralised in regard to these barbarous modes of slaying pregnant fish. In fact, in this particular, they retain the love of that kind of sport which is a characteristic of savages; and in explanation of their passion for it we are disposed to agree with Lord Kames, who refers the passion of some people for the chase to a remnant of the hunter's instinct, with which man was originally endowed, and which, though rendered dormant by civilisation, is not extinct, but specially active in certain individuals.

Salmon-fishing in Canada has been grievously injured by stake-nets, and especially by "burning the water," and the Indian's fish-spear, which Colonel Alexander pronounces a very ingenious implement. A prong at the end of a pole transfixes the fish, whilst two semicircles of wood on each side of the prong embrace its flanks, and enable the fisherman to raise it from the bottom of the stream. By a most foolish tenderness for the liberties of the Indians, they are allowed to use the murderous spear; and as, like our home savages, they use it mercilessly, and most successfully, against pregnant fish, we do not wonder to be told that the number of salmon in the St Lawrence and its numerous tributaries is rapidly decreasing. We have lying before us a list of no less than sixty-two of the chief salmon and sea-trout fisheries of Lower Canada, twenty-two of which are Crown rivers, and now open to public sale. In case any of our readers should think of speculating in these strange waters, we extract two or three notices from the Government advertisement.

"Discharging into the St Lawrence.—Musquarro: bold, rapid river; affords fine salmon-fishing with fly; good net-fishery station. Great Natashquan: famous stream; salmon of finest kind, and numerous. Nipimewecanan: fairy-like stream; falls nine miles inside; exquisite fly-fishing.

This fairy-like region, with its unpronounceable name, has, we fear, an unusual complement of Indians, who care for nothing but hunting and fishing.

"I think," writes Colonel Alexander, "the native love of excitement in the chase has something to do with their pertinacious pursuit of salmon by spears and flambeaux. It is a passion among some of the bands; and I must admit the habit has peculiar fascinations, and to many it is strangely exciting. Nothing can exceed the wild excitement with which these men pursue it. The sombre night-scene of the forest river seems to delight them. The elder man occupies the stern of the canoe, while the younger takes the 'post of honour' forward. The murmur of waterfalls and rapids drowns their exclamatory *ughs*, and the frequent splash that would else disturb the pervading stillness. With steady, stealthy speed, the birchen boat enters the rapid, and, cutting through its white waters, glides smoothly over the fall and into the 'tail' of the pool above, or across the quiet 'reach.' The blazing torch, stuck in a cleft stake, and leaning over the bow of the canoe, glares with dazzling brightness. The flame and shadow, swayed by ripples, conceal the spearers' forms, and bewilder the doomed salmon. Like moths, they sidle towards the fatal light. Their silvery sides and amber-coloured eyeballs glisten through the rippling water. The dilated eyes, the expanding nostrils, and compressed lips of the swarthy canoe-men, fitly picture their eager, excited mood. A quick deadly aim, a sudden violent swirl, and some convulsive struggles tell the rest. The aquatic captive, with blood, and spawn, and slime, and entrails, besmears the inside of the canoe. During a single night, from fifty to two hundred salmon may be thus slaughtered, and have as many more lacerated in their efforts to escape—the pools at such seasons being too shallow to afford certain safety in retreat. The speared salmon are sold to traders at their own price, as the deteriorating mode of capture so much depreciates the fish. That the Indians must suffer starvation by being deprived of their 'native liberty' to ruin our salmon-fisheries, is a very flimsy apology on the part of those who still desire to perpetuate so flagrant an abuse."

So, then, the same unreflecting spirit of destructiveness against the noble salmon is active alike in the confluents of the St Lawrence and the Tweed. Let all sportsmen, let all lovers of salmon, unite in putting an end to such disgusting barbarity. If consumers only saw a few specimens of unseasonable salmon struck by the spear, they would remember the loathsome sight, and take especial care to eschew a kind of food so unwholesome. Proprietors of salmon rivers are interested in putting down this cruelly destructive mode of fishing, for a reason with which Colonel Alexander's book first made us acquainted. There are few things, it appears, about which fishermen ought to be more careful than allowing their servants to clean the fish they have killed in the stream, or to throw their offal into it; for it is a well-known fact that the slightest tinge of blood, or the smallest portion of intestine, will alarm a whole shoal of salmon, and send them back in terror to the sea. The servants of the Hudson's Bay Company are well aware of this; and at all their fishing-stations, the place at which they clean the fish is at some distance from the river, and they invariably dig a hole in which they scrupulously deposit all the offal. We have a similar fact recorded in reference to the cod. When the fishing-grounds are fouled by garbage the fish invariably desert them.

We readily believe that salmon may be thus driven in disgust from rivers polluted with the fetid remains of their own species, just as we should shrink from dwelling or eating close to relics of human mortality; and we see no reason why the feelings of fish should not be respected by those desirous that they shall multiply abundantly in our rivers. And as salmon in the Tay may be presumed to be actuated by the same dislikes evinced by those within the territories of the Hudson's Bay Company, we think such persons should pray for an addition to that part of the Lord Advocate's bill which makes it penal to introduce into any river hot limes, refuse of gas-works or products thereof, prussiate of potash, or water in which green flax has been steeped, or sawdust, or refuse from a mill or manufactory or other work, or any matter or thing which shall poison or kill any salmon or trout.

But when this bill has passed, who shall describe the horror of the dirty drabs, mulcted in not less than one pound, and not exceeding two pounds, for placing in, or allowing to fall into, any river, at or below high-water mark, any coal, cinders, coal-ashes, dirt, or rubbish? Why, the worship of the foul goddess Cloacina will become extinct along the margins of our rivers and most odorous streams! The Lord Advocate, the first time he sails down the Clyde or walks by the Water of Leith, will assuredly be deafened by the applause of the refreshed divinities presiding over streams and rivers.

Seriously speaking, if this clause be enforced, it will not only promote the breed of salmon, but add immensely to the health and longevity of human beings, at present rendered miserable by dwelling near some great river, so polluted as to be, like the Thames, little better than a huge tidal sewer. Such a clause will preserve the Tay from ever being a gigantic nuisance like the Thames. And when all sorts of precious manures, now thoughtlessly thrown into rivers, must be stowed on land, who shall calculate the benefits to agriculture, when all animal and vegetable *débris* is consigned to the bosom of mother earth?

SALMON-REARING AT STORMONTFIELD, AND FISH CULTURE.*

WE have endeavoured to attract attention to the artificial rearing of fish, and especially of salmon. We had frequent occasion to refer to the interesting experiment in salmon-rearing at Stormontfield, about five miles from Perth, on the banks of the Tay. We have now the satisfaction of being able to refer all interested in pisciculture to an account of the manner in which it is being carried out, and of the practical results. Mr Brown, the author of the little work we refer to, has long devoted attention to the natural history of the salmon, stimulated, he informs us, by the experiments of Mr Shaw of Drumlanrig, from 1833 to 1838. Believing in the common opinion that the parr was a fish *sui generis*, which never attained a larger size than seven or eight inches, Mr Brown resolved to put Mr Shaw's statements to the test. Having, in February 1836, caught eigh-

---

* 'The Natural History of the Salmon, as ascertained by the recent Experiments in the Artificial Spawning and Hatching of the Ova and Rearing of the Fry at Stormontfield, on the Tay.' By William Brown, Perth. Glasgow: Thomas Murray and Son. 1862.

'Fish Culture: A Practical Guide to the Modern System of Breeding and Rearing Fish.' By Francis Francis. With numerous Illustrations. London: Routledge. 1863.

teen parrs in the Tay, he kept them confined in a stream of running water. By May the whole of them had become smolts; and so desirous were some of them to find their way to the sea, that they leaped out of their confinement, and were found dead on the bank. This having convinced him that the parr was indeed the young salmon before its assumption of the silvery scales of the smolt, he entered with earnestness into the artificial-propagation scheme when started at Stormontfield, and was present and assisted at most of the manipulations. His little work is most trustworthy, and deserving to be carefully perused by all interested in the natural history of the salmon, or desirous to understand the artificial mode of rearing it. The light which the Stormontfield experiment has thrown upon many obscure points connected with salmon is so important, and its value in demonstrating the commercial worth of salmon-rearing is so great, as to have attracted the attention of illustrious naturalists at home and abroad. Stormontfield has been inspected by visitors from France and Spain. Mr Brown states that the proprietors of the Tay fisheries were induced to undertake the artificial rearing of salmon in consequence of the representations of the late Dr Esdaile. This is true; but we happen to know that he constantly deplored the contracted scale of their operations, and insisted that, if intended to pay, salmon-rearing should be by millions, and not by five or six hundred thousand a-year. Only £500 has been expended on 300 boxes, laid out in parallel rows, with a path between each, 12 boxes being in the row, each box 6 feet long by 1 foot 6 inches, and 1 foot deep. The fall from the upper to the lower end of each box is 2 inches, and 2 feet in all, so as to allow the water to flow freely through them; but experience has proved that the fall is not sufficient, and that the quantity of water discharged from the feeding-pipe is too small. From these boxes the fry drop into a lower canal, where they are regularly fed with minced sheep's liver. About the end of May they are allowed to enter the rearing pond, which, as Mr Brown maintains, is the most faulty part of the whole plan, being only 223 feet long by 112 feet wide at its broadest part, and far too small to contain and nourish 300,000 fry, which the boxes can hatch: it would require to be four times as large. A second pond is indispensably necessary in order that the breeding may be carried on every year. At present, ova are deposited only every second year; because, if the fry of this year were introduced into the pond along with those of last year, the younger fish would be preyed on by the older.

Seeing that the experiment has been so successful as to demonstrate that to be amply remunerative it only requires to be extended, we are at a loss to understand the apathy of the Tay fishing proprietors. Rentals have largely increased of late years, and they seem to be satisfied; but their doings are not deserving of imitation. Wherever salmon-rearing is attempted, let it be on a scale which will yield important results.

We know of only one instance in which the artificial rearing of salmon has been prosecuted thus. Mr Thomas Ashworth of Poynton, Cheshire, having, along with his brother Mr Edmund Ashworth, purchased the Galway salmon-fishing, extending from Loch Corrib to the sea, resolved in 1852 there to try an experiment in the artificial propagation of the salmon. At the meeting of the British Association for the Promotion of Science, which met at Glasgow in 1855, Mr Edmund Ashworth read a paper, in which he stated that the Stormontfield experiment had demonstrated the practicability of rearing salmon of marketable value within twenty months from the deposition of the ova. "We are glad to learn," observes Mr Brown, "that Mr Ashworth is still carrying on his experiments, as this season he has deposited at the least 659,000 salmon ova in the tributary streams of Lochs Mask and Corra, where salmon ova were never seen before, although both these lochs are connected with Loch Corrib, which abounds with salmon—a natural barrier of rocks preventing the ascent of the fish. This barrier Mr Ashworth is about to remove, as not a single fish has been, observed to pass up. We have no doubt, however, but that the smolts reared in the upper tributaries will seek their way back to their native stream, and in a few years stock these lochs and rivers with this fine fish. Whether this gigantic experiment succeeds or proves a failure, Mr Ashworth deserves the thanks of the country for his labours."

We hope that more Scottish proprietors may be induced to follow his enterprising example; and as Dr Esdaile's recommendation induced the fishing proprietors of the Tay to engage in pisciculture, we trust our suggestion may have like effect in introducing it on the North and South Esk in Forfarshire. On the banks of both of these streams salmon might be reared much more profitably than at Stormontfield. The smolts leaving the rearing-pond there have to pass through more than 20 miles of the river before reaching the sea; and during this perilous descent their numbers must be frightfully diminished. But on the banks of these Forfarshire rivers, localities for salmon-rearing may be found so near the sea that in an hour or two after their release from the rearing-ponds the smolts may arrive at their feeding-ground in the ocean.

We are persuaded that in many localities it is practicable to raise salmon from the ova to the grilse state while the fish are in confinement. This is to be effected by having two ponds in proximity to each other, the one being fresh and the other salt water. Mr Brown refers to two such ponds erected at great expense at Stonehaven on the water of Cowie, belonging to the heirs of the late Alexander Baird, Esq. These ponds were originally intended for the purpose of hatching and rearing salmon from the ova up to maturity; but so carelessly were they managed that the result was a failure. Multitudes of parrs were produced in the little stream connected with the fresh-water pond, but as there was nothing to prevent them descending to the salt water, most of them, no doubt, did so, and met their fate. From personal observation Mr Brown asserts that until the parr assumes the smolt scales *it cannot live in salt water*. The following remarks should be considered by those wiseacres who have declared their belief that salmon spawn in the sea. "This fact was put to the test by placing some parrs in salt water. Immediately on being immersed in it the fish appeared distressed, the fins standing stiff out, the parr marks becoming a brilliant ultra-marine colour, and the belly and sides of a bright orange. The water was often renewed, but they all died, the last living nearly five hours. After being an hour in the salt water they appeared very weak and unable to rise from the bottom of the vessel which contained them, the body of the fish swelling to a considerable extent. We have taken ova which had been previously manipulated upon, and dropped them into sea water, which destroyed them almost instantaneously. We have put smolts, which have had on the scales for some time, into salt water directly from the fresh, and they seemed in their true element. These facts prove, that until the parr is covered with the new scales it is unable to live in salt water, and also that salmon cannot hatch in the sea."

Mr Brown, selecting five properly-developed smolts reared at Stormontfield, conveyed them per rail to Stonehaven, about sixty miles, and placed them safely in the salt-water pond on the 30th June 1860. For a few minutes they swam slowly among the brackish water at the side, but soon darted into the sea-water. On the 18th August Mr Brown saw three of his fish grown double the size. On the 15th April 1861, one of them was again seen, and reported to have doubled its size. The result was annoying. One morning a notorious poacher had been seen whipping the pond with his flies, and no more was seen of the fish. This case, no doubt, proves that the smolt will grow more rapidly in salt water than in fresh; but as none of these fish noticed by Mr Brown is estimated to have attained a weight of more than 1 lb. in ten months, the confinement, or the limited supply of food, evidently stinted its growth. A smolt of about an ounce, leaving Stormontfield, returns to the Tay in about six weeks weighing 3 or 4 lb. If salmon are to be reared for the market in sea-ponds, they must be artificially fed.

Supposing the poacher caught with a fly-hook these young salmon at Stonehaven, we have here corroboration of recent statements as to the capture of salmon at sea by means of the artificial fly. As it is so difficult to procure permission to fish for salmon in a good river, we counsel the lovers of the angle to try their luck in the open sea, where, we presume, they will not be interfered with, and may enjoy their sport undeterred by the pains and penalties of recent legislation for the protection of salmon.

To those who wish to multiply this noble fish in our Scottish rivers, Mr Brown's little work will supply valuable information. The art of artificially impregnating is easily acquired, and is thus described:—" So soon as a pair of suitable fish were captured, the ova of the female were immediately discharged into a tub, one-fourth full of water, by a gentle pressure of the hands from the thorax downwards. The milt of the male was ejected in a similar manner, and the contents of the tub stirred with the hand. After the lapse of a minute, the water was poured off, with the exception of sufficient to keep the ova submerged—this must always be attended to, even when the ova or milt is flowing from the fish— and fresh water supplied in its place. This also was poured off, and fresh substituted, previous to removing the impregnated spawn to the boxes prepared for its reception. We observed in this the first manipulation, and in all the others afterwards, that a very small quantity of the milt was sufficient to impregnate the ova of a large salmon, and that always a few of the ova, after receiving the milt, turned white; these were injured, and would prove addled. We also noticed that the salmon colour of the ova was heightened when the milt came in contact with it. Round tin pans, with as much water in them as covered the ova, were used to carry them to the hatching-boxes. The spawned fish were returned to the river, and went away after the operation quite lively." The curious fact is recorded that one of the male salmon manipulated on this the first occasion of artificial impregnation on the Tay, in 1853, having been marked, was recaptured, and manipulated again on its return to spawn in the same place in 1855.

We find no light thrown upon the perplexing anomaly first noticed at Stormontfield—namely, that about one-half of the fry in twelve months assume the smolt aspect, and migrate to the ocean, while the remaining half continue parr, and will not leave the pond till a year older. It was surmised, that remaining for two years might be a peculiarity of the progeny of grilses, and that the fry migrating when a year old were the offspring of fully-matured salmon. A conclusive experiment has demonstrated the futility of this hypothesis. The anomaly extends to the progeny of salmon as well as of grilses. "Why," observed Mr Brown, "those that remain behind for another year do so, we cannot tell, but such is the fact; and the best reason we can give is, that by this means the river has always fish in it that will migrate at least a month sooner in the spring than the fry of the first year; and also that male parrs will always be at hand in the river during the spawning months, in a fit condition to supply the want of male salmon when that occurs, which is a wise provision in nature, as many females in small and distant tributaries might be left without a mate if there were no parrs—male parrs having been proved to be in a breeding state at that time."

DIFFICULTIES SOLVED.

When Mr Shaw first announced the astonishingly precocious procreative power of the male parr, and asserted that by this means he had impregnated the ova of a large salmon, many doubted, and some pronounced it an impossibility. But this abnormal precocity, characterising only the male parr, has been also demonstrated at Stormontfield.

As might have been anticipated, another of the experiments there tried has given new proof of the ignorance of those who maintain that the ova of the salmon are impregnated, not by an emission of semen, but by actual coition.

Having denounced the notion of Mr Mackenzie of Ardross and Dundonnell that the salmon and the grilse are fish of different species, we received a communication from that gentleman reiterating his theory, and requesting replies to these queries:—" 1. A salmon of 6 lb., taken in the Tweed or the Tay 1861, when was he a grilse, and what was his probable weight as such ? 2. Where did the salmon that supplied our fisheries, when it was legal to catch them in November, December, and January, come from, seeing that all the grilses are at that time of the year in the rivers as kelts?"

The experiment at Stormontfield furnishes a solution of these difficulties, by proving that all the smolts do not return the same season as grilse, but that not a few of them remain in the sea, and do not return to the river until the spring or summer of next season, not as grilse, but as salmon of from 4 to 10 lb. weight. This is the explanation of small spring fish, and also of clean salmon coming to the rivers in winter. In both cases the fish had not returned as grilses, but as salmon.

One of the difficulties felt at Stormontfield was to discover a satisfactory means of so marking the pond fish that they might certainly be recognised when caught. The first expedient, that of cutting off the dead or second dorsal fin, has alone been found satisfactory. Many hundreds had a silver ring attached to the fleshy part of the tail, while others were marked with gilt copper wire; but none of them were ever heard of. It might be thought a dangerous expedient to punch a triangular hole in the gill-cover of so small a fish as a smolt. But this mode of marking, though found to be safe, did not prove permanent. The portion removed was speedily filled up. And now the Stormontfield mark is the absence of the dead fin, which cannot be simulated by those claiming the reward offered for the identification of artificially reared salmon. A French naturalist has hit on the ingenious expedient of introducing madder into the food of salmon-fry. Their bones being coloured by it, they are certainly recognised as artificially reared fish.

These experiments also show how small is the risk of artificial fish-rearing. In no year has there been a mortality of ten per cent. How different is it with salmon ova exposed to various accidents, and preyed upon by numerous enemies! The May-fly (*ephemera*) in its larva state is most destructive. "One year," Messrs Ashworth state, "we deposited 70,000 salmon in a small pure stream adjoining to a plantation of fir trees, and these ova we found to be entirely destroyed by the larvæ of the May-fly, which, in their mature state, become the favourite food of smolts or young salmon. We know that the natural enemies of the salmon cannot be destroyed, as they exist both in rivers and in the sea; consequently there is left but one certain mode of increasing the quantity of salmon, and that is by artificial means—by collecting the spawn, and placing it beyond the reach of its enemies for hatching and protection for the first year of its existence; and this may be done in vast quantities at a small cost, and without injury to the parent fish."

We pray those meditating how to encourage the breed of salmon to remember that these are the words, not of sanguine schemers, but of sagacious English gentlemen of high position in Lancashire, and perfectly able to appreciate the capabilities of pisciculture as a commercial speculation. We have no doubt that their fishing property in Galway will be managed so liberally as to yield important results, and put to shame the want of enterprise displayed by the fishing proprietors of the Tay. The experiment at Stormontfield, besides its scientific value, has in a pecuniary view been very satisfactory. "We have no doubt about the matter," observes Mr Brown, "for on referring to a statement of the rental of the Tay, published by the proprietors themselves, we find that in the year 1828—the year of the passing of Home Drummond's Act—the rental was £14,574. It gradually fell every year afterwards until 1852, when it reached the minimum of £7973. In 1853 the artificial rearing commenced, and in 1858 the rental was £11,487. It has now, 1862, reached what it was in 1858. We are aware that other reasons are given for the rise in the rental, such as the extra price of the fish in the London market; but we should like to know how it happens that all the other rivers in Scotland (with the exception, perhaps, of the Sutherland rivers, which have the same market for their fish) have, since 1852, had a lower rental instead of a higher." Even after making allowance for the benefit derivable from the earlier closing of the Tay during the years 1853-4-5, which was legalised by an Act passed in 1858, Mr Brown is "of opinion that the great rise in the rental in nine years, cannot be accounted for in any other way than from the pond-bred fish ; and if the fishing proprietors would see to their own interests, they would have many acres of breeding boxes and ponds made to rear and preserve their fish. At present, however, their rental is in the ascendant, and they are contented; but should a reverse take place, we would then see artificial propagation much in favour." We cannot conclude without beseeching them to get rid of the narrow-minded feelings which have induced them to refuse the request of those soliciting a supply of salmon ova for the purpose of artificial propagation in foreign countries. Garibaldi met with a direct refusal; and compliance with an application from the French Government was evaded! Our Scottish rivers might with great advantage receive the ova of various species of foreign fish. But how shall we preserve them, if we anger their possessors by such pitiful displays of a lack of international politeness? This want of courtesy is in marked contrast to the liberality of the French Government, in dispersing ova from their rearing establishment at Huningue. Besides despatching annually millions of ova to the various rivers of France, large quantities are also sent abroad to various parts of Europe.

This establishment, near the Rhine and Rhone Canal, covers a space of about seventy acres, and is conducted at an annual cost of £2200. It is calculated that twelve live fish are produced for a penny.

Mr Francis, whose 'Fish Culture' we would commend to the study of those interested in its subject, bitterly laments the little attention which is bestowed upon our national fisheries, and eloquently descants upon their capabilities, especially when aided by the resources of pisciculture—" a science the workings and effects of which have to be discovered, and in which there are honours and profitable fame to be won." He is sanguine that where the best species of fish can be cultivated, water may yield a revenue very far exceeding that of land. He asks what edible land animal possesses the grilse's capability of reproducing 9000 of its own size in less than two years? Easy and inexpensive though pisciculture be, he observes that many would willingly practise it, if they knew how to set about it; but they know not where to obtain fish spawn; nor can they find persons sufficiently instructed to take spawn from the fish, or to set up apparatus for them, because there are probably not half-a-dozen people in this country who know how to do so. We agree with him that in a country like ours, so fitted by nature for fish-rearing, a government establishment, like that at Huningue, is eminently required as a central depôt for the supply of fecundated ova, and for promulgating practical instruction how to rear them. But being hopeless that Mr Gladstone will crave the Commons to grant a subsidy for such a purpose, we must comfort ourselves with trusting that the perusal of such books as those of Mr Brown and Mr Francis will stimulate many to carry out by personal enterprise the much-needed work of stocking our lakes and rivers with the more valuable kinds of fish.

Water-culture, in his opinion, is a science yet in its infancy. " We should know what kind of food suits our various fish best, and what conditions best produce that food, and how these conditions are best to be cultivated, so that such food may be self-producing." This chapter on the food of fish and its production, demonstrates, in a very interesting manner, how great is our ignorance of this department of knowledge. " There is not an insect or small reptile that inhabits the soil beneath us, or the waters around us, that is not food for fishes in a greater or less degree. Worms of all kinds, flies, grubs, larvæ, cockchafers, crickets, leeches, snails, humble-bees, young birds, mice, rats, all serve the turn of one fish or another, and so in turn help to produce food for man. Nothing living comes amiss, but doubtless some kinds of food agree with them far better than others. But we know very little on this branch of the subject. It is dreamland to us, with a very little waking reality."

Some rivers produce larger trout than others, like them in all visible features. Mr Francis compares the trout of the Chess, a branch of the Buckinghamshire Colne, with those of the Wick, little more than a good-sized brook, near High Wycombe. The May-fly abounds on the Chess, but is hardly seen on the Wick, and yet the Wycombe fish, attaining a size of from seven up even to ten pounds, and of a red colour deeper than salmon, is decidedly superior. Why? How is it that this little stream has not its equal in all England for the size and flavour of its trout? Mr Francis surmises that the fresh-water pulex or screw has not a little to do with the production of this marked superiority. " I have seen the trout picking them at the walls, which pen the stream in some places, as rapidly as a child would pick blackberries from a hedge; and I am induced to think that this insect has much to do with the fineness of the fish; and the more so because, when I have found it to exist in any quantity, I have invariably observed that the trout are of fine size, and in unusually good condition."

Hence the feasible suggestion that the entomologist and the botanist shall assist pisciculture by ascertaining the kinds of living creatures and plants on which fish thrive most, and by pointing out how they may be introduced into localities where naturally they do not abound.

The bottom of Loch Leven is in some places covered with a peculiar weed, sheltering various insects, chiefly crustaceæ, and small snails of various sorts; the lake also abounds in the more minute entomostracæ. Large quantities of both are found in the stomachs of the trout. And therefore, when recently consulted as to stocking a certain water with this prized trout, we insisted on the ova being accompanied by a quantity of the weeds and small stones to which the favourite food of the fish would adhere.

Besides noting what fish eat, Mr Francis would have pisciculturists note the creatures that eat fish or their spawn. He is particularly wrathful with Thames swans, whose voracity for fish spawn he has often witnessed. Allowing a swan the moderate daily allowance of a quart of trout spawn (containing at least 50,000 eggs), and supposing that only 200 swans are at work (about a fourth perhaps of the number really employed), he makes out that in a fortnight they will have devoured a hundred and forty millions—a heavy price truly to pay for the picturesque appearance of these dire enemies of fish-culture. Knowing the mischief they do, Mr Francis detests swans, and prays that every swan may, like the uncommon black swan, become a *rara avis in terris*, or rather *aquis*. He suggests, moreover, to the lovers of the picturesque, that a couple of stuffed swans made mechanically to bend their necks now and then, and anchored at some distance from the banks, will be a very comely and decidedly safe addition to the beauties of a river devoted to the culture of fish.

SOMETHING MORE ABOUT THE HATCHING OF FISH.*

WE have done something, we hope, to interest our readers in the very important question, How is animal food to be provided for the rapidly-augmenting population of the British Islands? We have pointed out that, in the ocean by which they are surrounded, and in the lakes, rivers, and canals by which they are intersected, bountiful nature has provided for all her children alimentary substances of the greatest value, and to an extent which, without exaggeration, may be pronounced inexhaustible. Let human beings be multiplied in such myriads as to horrify Malthusians, our "cheerful faith" is that the great Father has made the fish-pond of the seas and rivers bear such a proportion to the families to which He has given "the dry land" as their dwelling and harvest-field, that food shall not fail, provided advancing intelligence and industry be unceasingly applied to the solution of the increasingly intricate problems of our social life. This *proviso* being attended to, we may hope for something better than was prayed for by the benevolent King of France, who wished that

* 'Fish Hatching.' By Frank T. Buckland, M.A., M.R.C.S., F.Z.S., Student of Christ's Church, Oxford, and late Assistant-Surgeon, Second Regiment of Life Guards. London: Tinsley Brothers. 1863.

every peasant might have a fowl in his pot. We are sanguine that salmon shall again smoke on many a farmer's table, and even find its way to the kitchen and the bothy; that oysters shall be multiplied countlessly on many a depopulated scalp; that superior species of fish shall swarm in many of our streams; and that our population generally shall become ichthyophagous to an extent as yet unknown.

Does any one ask a reason for this faith in the increase of fishes? Did we not read, in an Edinburgh newspaper, an advertisement headed "Fish *versus* Butcher-meat," with the pleasant notification, that cod and skate, haddock and halibut, might be had at from 2d. to 3d. a-pound? Did we not read of the great joy in a fishing village on account of the capture of a skate so huge as to be a burden to three lusty fishermen, though this be a fish generally of small repute in Scotland? And, greatest marvel of all, have not her Majesty's faithful Commons benignantly thought of the multitudes living on short commons, opposed the Government, and victoriously supported Mr Fenwick's motion for a commission to inquire into the means of improving the sea-fisheries of Great Britain and Ireland? In spite of those who think that the end of the world is at hand, we really believe, with Galileo, that it *is* moving, and in the right direction too —not to perdition, but in the divinely-appointed path of procreation and progress. The mission of human beings is to people the earth. In fulfilling it they must learn to reap crops alike on land and water: in both seed must be cast. And if we have not lost all faith in Providence, and in the assertions of philosophers seriously addressing themselves to the question how food is to be found commensurate with the increase of the world's population, we may confidently expect an ample reward to the labours of *aquæculture*.

"We are," says Mr Buckland, "for the most part fully cognisant of the inhabitants of the land, but how little do we know of the inhabitants of the water! Man has dominion given him over both land and water. Of the former he has taken every advantage; from the earliest times there have been *agri*culturists, or land farmers. Who ever heard of an *aquæ*culturist, or water farmer?" We have; and Mr Buckland might have heard of him too, if he had chanced to read a remarkable letter of Mr O'Ryan de Acuna, published in 'The Field,' August 8, 1857, by Mr Thomas Ashworth, to whom it is addressed:—

"In lieu of pisciculture, which would only include *vertebratæ*, I shall continue" (writes this gentleman, who has obtained the privilege of bringing under cultivation the waters of Spain) "to make use of the word '*aquæculture*,' as including within its signification the other orders of zoology, the cultivation of which, in molluscæ and zoophytes, has also been recently practised on principles of sound philosophical induction.

"Aquæculture has certainly not obtained, up to the present day, that profound attention which its vast importance entitles it to. We are in the habit of seeing the earth being cultivated, but our ideas have not yet been awakened to the propriety of cultivating the water likewise. Everybody knows there is a period in the life of human societies, during which even the earth is not cultivated. In that period the spontaneous productions of the earth suffice to satisfy the necessities of the individuals composing such communities; but as soon as they are about to abandon the savage state, agriculture becomes necessary, and consequently it takes rise. The necessity of cultivating the waters, on the contrary, does not show itself before the moment in which communities have acquired a high degree of civilisation, at which time the application of aquæculture becomes no less indispensable than that of agriculture at the instant savage life had to be relinquished. No doubt the cultivation of the water becomes a necessity at a much later epoch than the cultivation of the earth;

but this arises, in a great measure, from the lesser control we possess over the waters, and equally so from their superior fecundity to that of the land; and such a superiority affords strong argument for the incalculable advantage that may accrue from an intelligent and ample system of cultivating the liquid surface of our planet. If we look rightly into the matter, it would appear absurd to deny that the time is not actually come for us to be obliged to bring the waters under cultivation."

Mr Buckland will, we doubt not, rejoice to learn that the water farmer has not only been heard of, but is already located through wider portions of the world than he (Mr B.) seems to be aware of. In the appendix to his very amusingly written as well as useful little book he gives, for the benefit of his inquiring readers, as complete a list as he can of works on pisciculture. It does not include that of M. Jourdier, published in 1856 for a couple of francs, and forming part of Hachette & Co.'s Railway Library. "This work," observes the distinguished M. Coste in a commendatory preface to the reader, "is of a double utility, being the most complete published up to this date, and comprehending both marine industry and the rearing of leeches." We now refer to M Jourdier's cheap and excellently illustrated work for the purpose of indicating to those in search of piscicultural knowledge where they may find a list of books far more extensive than that furnished by Mr Buckland. It fills more than two pages, and should be consulted by all pisciculturists able to read French.

We are glad to have been the means of diffusing much fish lore not accessible to the general reader, and only to be acquired by acquaintance with a new department of literature much cultivated abroad, and now becoming known in this country by communications to 'The Field,' and by the writings of the late Dr Esdaile, Mr Thomas Ashworth, Messrs Buist and Brown, Perth;

and more recently of various writers in England, among whom may be mentioned 'Ephemera, or the Book of the Salmon;' Mr Francis; and, latest of all, Mr Buckland, whose good-natured rollicking style pleases the general reader, while his extensive and accurate information satisfies the more scientific inquirer. We hail him as an ally, and claim the fulfilment of this promise—"The oyster must and *shall be* cultivated in this country. I propose shortly to take the matter in hand. M. Coste and the French pisciculturists have done so much that we ought to be ashamed of ourselves for being all behind-hand in this important matter."

In Scotland we have especial reason to look foolish and to be ashamed: in the management of our fisheries we are as great fools as our worst enemies could wish us to be. Well may Mr Buckland declare—

"We have been asleep—we have had gold nuggets under our noses, and have not stooped to pick them up. Tons of fish, worth thousands of pounds, only want a net placed round them to be converted into bank-notes; but they want looking after, they want cultivation. You must not kill your 'golden fish' (the 'golden goose' may now retire on half-pay); you must not watch the spawning fish-mother to her nest, nor must you permit others to do it—for the sake of her unwholesome carcass (for which the French cook at the Palais Royal will give you a franc or two) destroy her, and at the same time thousands of young fish.

" *O fortunatos nimium, sua si bona nôrint!* would dear old Virgil have said of the *aquæ*-culturist, if he had known what we now know. You must not, O friend, put your heel upon yon mass of tiny round balls, which, if, properly treated, would most assuredly in about four years develop themselves into huge, silver-coated salmon, and, what is more, will cost you not a penny for food or keep."

We demur to the statement that it may take "about four years" to develop a tiny smolt into "a huge salmon," if we are allowed to apply such an expression to a fifteen-pounder,—a " huge" increase, at all events, from about two grains, the weight of salmon-fry three days old. Mr Buckland forgets the following observations of Mr Ashworth of Cheadle, near Manchester :—

"The fry, at three days old, is about two grains in weight; at sixteen months old it has increased to two ounces, or 480 times its first weight; at twenty months old, after the smolt has been a few months in the sea, it has become a grilse of eight and a half pounds,—it has increased 68 times in three or four months; *at two years and eight months old it becomes a salmon* of twelve to fifteen pounds in weight; after which its increased rate of growth has not been ascertained; but by the time it becomes. thirty pounds in weight, it has increased 115,200 times the weight it was at first. I do not suppose there is any other animal that increases so rapidly and at so little cost, and that becomes such a valuable article of food."

The anomalous diversity of growth in young salmon, so very remarkably illustrated at Stormontfield, is, we think, still a mystery. In the office of the conservator of the Tay fishings were three specimens spawned from salmon-roe about the end of December 1861, hatched in April 1862, fed in the same pond, and yet exhibiting these peculiarities. The first weighs 646 grains; the second, 135 grains; the third, 26 grains. Our worthy old friend "Peter of the Pools," *alias* Mr Robert Buist, Perth, being desirous to have the opinion of a distinguished naturalist as to the cause of this strange anomaly, sent these fishlings to Mr Buckland, who thus replies:—" I submitted the specimens and letter of 'Peter of the Pools' to the scientific meeting of the Zoological Society. J. Gould, Esq., F.R.S., and Dr Günther of the British Museum, gave it as their opinion that, provided always the evidence of their being of the same age is well proved, this was simply a case of cause and effect—the bigger fish being the stronger and most healthy of the lot. I myself quite agree with this. A number of fish are turned out simultaneously into a pond; some are weak, some are strong; the stronger, of course, gain the mastery over their brethren, and gain all the advantages of the pond, whatever these may happen to be; the consequence is, that, in proportion to their advantages, they become larger than those which have them not. The same thing happens, so to say, in human ponds; for in large cities we find that the babies and young children who are well fed and live in good air are much stronger and healthier, and, for the most part, larger too, than those born and bred in crowded courts and back passages, and who feed on red herrings and tea rather than on butcher's meat and beer. Take a given number of children from a given large city, say a hundred of the same age, and put them side by side. I doubt not that we should be able to pick out three specimens from among them whose full-length photographs, if grouped together, shall show as much difference as do the drawings of the three fish now before the reader."

He then relates the result of a dissection by himself next day, which showed that the stomach of the smallest specimen "contained nothing, or positively next to nothing;" and the conclusion arrived at is, that the two larger specimens, whose stomachs were full, were indebted to differences of natural vigour, and also of food, for their remarkable size.

Now, with all deference to learned zoologists, we cannot accept this as a solution of the phenomenon which all along has puzzled the Stormontfield experimenters; and, though we have not had time to communicate with them, we venture to prognosticate that they are not satisfied with the philosophy of Mr Buckland, who argues, if we find a specimen of " a little old man," why should we not find "a little old fish?" We beg pardon for asking, Is not Mr Buckland a little of an odd fish? He writes as if the common diet of babies and young children in great cities were either red herrings and tea, or butcher's meat and beer. If it be so in England, we have an explanation of juvenile mortality, for the checking of which we advise "parritch." Moreover, as "Frank T. Buckland, late Assistant-surgeon Second Regiment of Life-Guards," may be reasonably supposed not to have had very ample opportunity of introducing babies into her Majesty's dominions, we should like to have the opinion of Dr Simpson of Edinburgh on this question. Three young salmon, known with certainty to be of the same age, and to have been supplied with the same kind of food, are found at the end of twelve months to be of the respective length, $6\frac{1}{2}$ inches, $3\frac{5}{8}$ inches, $2\frac{1}{8}$ inches. In any "human pond" with which you are acquainted—in Charlotte Square, Edinburgh, for example—do you know of like diversity among babies, born at the same time, of healthy parents, and fed in the same way? And is it consistent with your experience that, at the usual age, fifty per cent of female children attain the size of puberty, and the capacity to perform all its functions, but that other fifty per cent do not attain that size and capacity, and remain at home, saying, like the sweetheart of "My boy Tammie,"

"We're ower young to marry yet,
We canna leave our mammy"?

Until Dr Simpson answers these questions in the affirmative, we are confident that "Peter of the Pools," and his well-informed piscicultural coadjutor, Mr Brown, will think that a shallow explanation of a deep mystery has been all that they have got from the Zoological Society. The truth is, Mr Buckland and his learned friends have misapprehended the nature of the puzzle submitted for their solution. He writes as if the anomalous growth of salmon only amounted to three per cent; whereas it is nearer fifty! About a half of the Stormontfield fry do not at the end of the year exhibit the silvery lamination of the smolt, and refuse to leave the rearing-pond; while the other half, greater in size, and protected by a different sort of scales, insist on getting to the sea, from which, in about forty days, they begin to return as grilses. This anomaly cannot be explained, either by the " little-old-man" similitude of Mr Buckland, or by the difference-in-food theory of Mr Goold and Dr Günther. It is, we suspect, somehow connected with the recently discovered habit of the salmon species, which occasions a double or divided migration to the sea as smolts, and from the sea in their later stages. When this is better understood, we shall, doubtless, be furnished with new proof of Providential wisdom.

Mr Buckland, as well beseems a late assistant-surgeon, treats his readers to a "bit of anatomy."

"Here," says he, "is a preparation from a salmon, which shows that the ova are thrown off from a long finger-like membrane, one side of which is laminated like the leaves of an opened book; it is in these leaves that the ova are secreted, and some of them may be seen still adhering *in situ*.

"I have ascertained that behind the ova, ready to be extruded, say this year, are other ova, as small as pins' heads, which will arrive at maturity next year. When the ova are ripe, they detach themselves from the membrane, and lie quite loose in the cavity of the abdomen; they are not, however, I believe, all shed at the same moment, but at various intervals—so say observers of salmon-spawning. They say correctly, as it is not likely that all the ova should become loose at the same moment."

Attention to this ascertained fact must be paid by all prosecuting pisciculture by means of artificially impregnated ova. Excess of pressure endangers the parturient fish; and ova prematurely extruded by the manipulator are not in the condition for being impregnated by being mixed with the milt.

When impregnating the ova of trout from the Wandle, Mr Buckland remarked, for the first time, that the ova of some of the trout were of a splendid coral—red-coloured; others, on the contrary, were almost as white as peas,—yet all good eggs. This depends, it is said, upon whether the trout is red-fleshed or white-fleshed. And we subsequently ascertained that the young fish hatched from the red eggs were much brighter than those from the yellow. A correspondent informs him that the same great variety exists in the colour of grayling ova, though the parents in that fish are not red-fleshed; and also points out that, though we have no red-fleshed hens, there is a variety of colour in the eggs daily presented at the breakfast-table. He adds, "the pale and the red fish ova are equally fertile, and the colour does not depend on the age of the parents. These two points I have proved. I cannot believe it to depend on colour of flesh, and therefore attribute it solely to variety in feeding."

With this statement we were disposed to concur, as we constantly observe that when our fowls are not properly fed the yolks of their eggs are pale and flavourless. But we were quite at a loss to explain why the shells of our chittagong's eggs are pink, while those of our Spanish fowls are white. As they receive the same sort of food, the difference in the colour of the shells seems to indicate that this is a characteristic difference of species. And if it be so with the eggs of fowl, why may not difference in colour also indicate like difference among fish? Consequently, the first point to settle as to different-coloured fish eggs is, whether they have been taken from the same variety of fish.

The ova of fish are exceedingly hard and tough, and so elastic as to rebound from the floor like an india-rubber ball,—a beautiful provision to prevent them being crushed or injured by the stones amongst which they are deposited.

"I was much surprised," observes Mr Buckland, "to

find the eggs of the trout at such a considerable depth in the gravel, certainly from one to two feet. They were all about loose in the gravel, reminding one of plums in a pudding. I cannot understand how the trout manages to get her eggs so deep in the sand. They certainly sink in the water; but one would fancy the current would whip them away in a moment. Again, how are they not crushed? I have stated above that their coats are very elastic; but I had no idea they were so tough. In order to ascertain positively how much direct weight they would bear, I tried experiments on the eggs, by placing iron stamped weights on individual eggs. I was astonished to find that they were not crushed till I had placed no less than five pounds six ounces on them."

This elastic toughness facilitates the transport of fish ova for the purposes of pisciculture; and experience has demonstrated that they can be safely transported hundreds of miles, either by land or water, if carefully packed in layers of moist moss, or of rough sponge, the size of a walnut, and well cleaned. The boxes of ova sent to London from the famous fish-rearing establishment of Huningue, usually occupy from two to four days in the transit; which is safely effected in consequence of the French pisciculturists insisting on observance of this simple rule—*never attempt the removal of ova till the eyes of the fish are plainly seen in the egg.*

The expense of collecting and removing ova is trifling. The superintendent of Messrs Ashworth's Irish fishings collected and deposited seven hundred and seventy thousand salmon ova in the streams of Lough Mask, besides conveying alive forty adult salmon a distance of twenty-three miles in a large tub of water. The cost of this very laborious-looking operation was only *eighteen pounds*, in addition to the weekly cost of the staff of water-bailiffs and workmen.

We are delighted to learn that success has at last rewarded the repeated efforts to introduce the ova of

British salmon into the salmonless rivers of Australia, by transmitting them packed in ice. In March 1858, we wrote : " In order to prevent unnecessary pains to protect salmon ova from the effects of cold, it should be known that, if gradually applied, it does not injure them, and that they have been known to be frozen up in a sheet of ice without losing their vitality. Réaumur demonstrated that the eggs of many insects are equally uninjured by excessive cold." If this had been considered by those interested in the experiment of transmitting salmon ova to Australia, the trial of the effect of freezing ova might have been made years ago.

The temperature of fishes, as we stated in the same article, on the authority of Liebig,* is from $2.7°$ to $3.6°$ higher than that of the medium in which they live; and as a warning that their ova are far more apt to suffer from excess of heat than from excess of cold, we related the failure of the attempt, made at our suggestion, to rear young salmon at the Crystal Palace, Sydenham. Instead of being in a temperature of about $40°$ Fahr., the ova, sent from the Tay, were placed in the tropical department, contrary to our wishes, and could not endure a temperature of about $60°$. In like manner Mr Buckland's young fish sickened at the same degree of heat arising from the direct action of the sun, and were only saved by the prompt application of ice. And the London pisciculturists have now arrived at the conclusion that the young fry soon begin to pine if the thermometer marks many degrees over $55°$.

We have alluded to the inexpensiveness of the first step in the process of pisciculture—namely, collecting and depositing the eggs. The hatching apparatus, our readers will rejoice to learn, is neither costly nor cumbrous. That patented by Mr Boccius, and sold at the ridiculous price of £10, 10s., though capable of containing 25,000 salmon ova, is only two feet long by one broad, and requires not more than four inches depth of

* 'Letters on Chemistry,' p. 67.

water. Those used by Mr Smee, and exhibited at the 'Field' Office, 346 Strand, London, may be made either of zinc or earthenware, and are twenty inches long, four and a half deep, and six wide. They can be placed one above the other, like the steps of a stair, so that the water shall fall from the one to the other by means of lips. These boxes may contain either a series of glass rods, on which the eggs can sit, which is a neat and clean way of hatching them; or may be filled with half-inch gravel, *boiled*, in order to insure the destruction of pernicious insects and animalculæ. "By next September or October," says Mr Buckland, "I shall be able to tell the reader where these may be procured, or, anyhow, can give him a model." This is very obliging; but, without availing ourselves of his offer, we can readily supply ourselves with a cheap and effective hatching apparatus, either for the parlour window or the park. We only require to read any illustrated work on pisciculture,* in order to understand what is needed; and with our own hands, or the help of a carpenter, it may be readily constructed.

Mr Buckland writes very pleasantly regarding the enemies of fish while in the ova, or after being developed. The common house-rat finds the ova a *bonne bouche;* ducks must be driven off if we want fish, and the stately swan must be banished with them. Milbourne, the water-bailiff on the Thames, near London, speaking of swans, gives this graphic description of their doings: —"Lord bless you, sir! they not only *eat* the spawn, but they eat *nearly all of it.* The number of swans already between Walton and Staines is beyond belief. They swarm there; and if they're to be allowed to breed, we shall have such a mass of swans that the river will be regularly smothered with them. Suppose they don't know where nor how to find the spawn? Gammon! Don't a donkey know where to look for thistles, and don't I know where beefsteaks grow?"

* We recommend that of M. Jourdier, already referred to.

The water-ouzel has unexpectedly got justice done to him by the solemn verdict of a scientific meeting of the Zoological Society in February 1863. Having been fairly put upon his trial as a notour destroyer of fish-spawn, the first verdict was " Not proven." This being the form of a Scotch verdict, an English water-ouzel was entitled to enter a demurrer. We are therefore not surprised to learn that a distinguished ornithologist objected to it, and that the jury ultimately returned their verdict thus : " Water-ouzel *fully acquitted* of the charge of eating fish-spawn."

The fact is, the water-ouzel frequents the spawning-beds in order to prey upon the insects which feed on the ova of the fish ; so that in our merciless ignorance we have been killing a piscicultural friend every time we shot a water-ouzel ; just as, we believe, the farmer hangs a friend every time he traps a mole.

As to the feeding of " water-babies," as Mr Buckland affectionately calls his fishlings, they resemble land-babies in this, that they are more apt to suffer from too much than from too little food. The favourite fare at Stormontfield is pounded sheep's liver; but in France pounded fish is much used; and M. Jourdier thus writes of a kind of diet which we commended to public approbation in " Hippophagy, or should we eat our Horses ? "—" Nothing is easier than to feed young fish with the muscular flesh of the ox, bruised or cut so as to suit the size of the little creatures ; and if this be too dear, its place may be supplied by *horse-flesh* cooked or raw, dried and pulverised."

THE HERRING.*

"Voluntary poverty" may be a merit in the eyes of narrow-minded religionists, who have not sense enough to know that God has given us all things richly to enjoy: for a nation to be poor, by neglecting to avail itself of the bounties of Providence, is political suicide, accomplished by a lingering agony, to witness which is pitiful. The wretchedness of Ireland was a scandal to Europe, until the recent dawn of brighter days for that unhappy land; and however proud of our country, we Scotch must sigh and blush when we think of the miserable penury which is chronic in certain Highland and maritime districts, and which elsewhere recurs periodically with a severity necessitating spasmodic efforts of charity, frowned upon by political economists, and not approved of by the reason of the benevolent, who follow the impulse of feeling even while their judgment is not blind to the comparative uselessness of a remedy known to be but temporary.

Until remunerative industry is provided for those able to work, charity is a makeshift which aggravates the evil it can only palliate for a season. The astonishing fact in connection with the penury of certain portions of the Brit-

* 'The Herring: Its Natural History and National Importance.' By John M. Mitchell, F.R.S.S.A., F.S.A.S., F.R.P.S. With Illustrations. Edinburgh: Edmonston & Douglas. 1864.

ish people is this—employment unfailing, and sufficient to supply all they need, is at their doors, but they will not resort to it. The Celtic race seem affected with somewhat of the horror of the sea which characterises the modern Hindoo, and with that aversion to fish and fishing exhibited during the uncivilised period of ancient Greece.

A writer in the 'Quarterly Review,' in a notice of Mr Yarrell's 'British Fishes,' says of the starving Irish, "These people have their salvation before their eyes, but they will not turn to it with a good heart." " It is the same," continues this writer, "or even worse, with the Hebrides at this moment. We happen to number among the most esteemed of our personal friends one of the principal proprietors of that interesting archipelago; and we are assured that though, during thirty years past, that family has made every effort to encourage sea-fishing among their dependants, it has never been in their power to procure, except during the smoothest weather of summer and autumn, a decent supply of sea-fish even for their own table."

This is corroborated by the recent statements of Mr Andrews, President of the Natural History Society of Dublin, who, when speaking of the depression of the Irish fisheries, says that the salt fish imported into Ireland annually amounts to 1200 tons, valued at £27,000; and that the annual import of herrings is about 80,000 barrels, valued at £128,000; and this, too, at a time when the Irish fisheries, if properly worked, are far more than equal to any demand which can be made upon them. Irishmen living on imported fish! this is really too absurd. The provoking part of this Irish folly is yet to be told. Enterprising foreign craft come to the coast of Ireland and carry off the treasures of the deep before the very eyes of the often-starving natives of the country. Hearing such things makes one ask, Of what use, then, is the sea?

Such statements may surprise some of our readers, especially those living in large towns, fish-fed by the cheap and rapid means of transit furnished by the railway; and those not familiar with the household economy of "huts where poor men lie" may fancy that ichthyophagy must be far from rare. And it must be owned that certain things seem to indicate that such is the fact. For instance, in the very interesting evidence given before the Royal Commission for visiting the Universities of Scotland by the late Principal Lee, we find the reverend gentleman declaring that, when a student at Edinburgh College, he knew young men living for a session chiefly on herring and oatmeal. And another reverend doctor, an intimate friend of the Principal, has often laughingly declined to taste herring at our table, because he ate so many when a student that he was satisfied for life! Moreover, our Scottish literature has about it an undeniable smack of the herring. Our daily talk, our moralising proverbs, our love-songs, have all a savour of this fish. "Dead as a herring" is said of one who is "gone" beyond remeid, and is probably an allusion to the fact that the herring dies the moment that it is taken out of the water, in consequence of the gill-covers being very loose and opening wide. And when we wish to indicate a man's individual responsibility, we tell him significantly "ilka herrin' man hing by its ain head." When the Scottish swain seeks to ingratiate himself with his sweetheart, he brings before her a picture of the "gear," to the worth of which he flatters himself she will not be blind. Distrustful of the value to be attached to "a hen wi' a happety leg," he adds the item—

"I hae laid a herrin' in saut;"

and then ventures to ask—

"Lass, gin ye loo me, tell me noo!"

Were any one to quote sundry passages from Burns to silence the hypocrite who should pretend that the

Scotch don't like whisky, he would certainly *nonplus* his antagonist. How, then, after being so frank in our admission of the fishy odour pervading Scottish life and conversation, can we maintain that the Scotch neither catch nor eat enough of herring?

As to the poor students at Edinburgh University living on herrings and oatmeal, that is not to be wondered at, considering the proximity of Newhaven to the Scottish metropolis. But since the days of Principal Lee we suspect there is no small change in the style of student diet. At all events, when attending that renowned university, we certify that our fellow-students, though daily hearing the (now, alas! discontinued) merry chimes of St Giles's playing "Caller Herrin," responded to the call,

"Wha'll buy my herrin', fresh frae the sea?"

in a fashion not likely to find favour with a fishwife. They were addicted to having a *red* herring to supper, having made the curious physiological discovery that the fish so prepared, if eaten and washed down by a pot of porter, left on the palate so strong a flavour of raw oysters as to necessitate its removal by the speedy imbition of a tumbler of toddy. As to the inference regarding the fish-eating habits of the people which a stranger might draw from the ichthyological similitudes to be found in Scottish songs and proverbs, we must correct it by means of our better knowledge. We are not even now a fish-eating race to anything like the extent which might be anticipated from our maritime position. We do not eat salmon because they are too dear, and must so continue until pisciculture be resorted to on the large scale which we have so often recommended; cod and haddocks are not habitually eaten even by the middle classes; skate, generally undervalued, is scorned by the multitude; and as to eels, the antipathy to them is notorious.

As to the herring, the extent to which it is captured

is very great. Still we are not sufficiently alive to the national importance of this branch of marine industry. The ignorance of most of us in regard to the natural history of this most interesting fish is astonishing; for which, indeed, we are not altogether to be blamed. Our "practical men," and our reputedly learned men besides, have been too long blind leaders of the blind. Assertions have been made and assented to, in regard to the habits and the habitat of the herring, which ought to have been discontinued and disbelieved long ere now. But as ignorance in regard to this important fish, and partial comprehension of its influence on our national fortunes, are still prevalent, we are glad that the publication of Mr Mitchell's very useful work affords us the opportunity of treating of various matters connected with this branch of British fisheries.

This gentleman, occupying the position of Belgian Consul at Leith, and living on the banks of the Forth, has excellent opportunities of examining into the natural history of the herring and its congeners; and the information he communicates is the more trustworthy, from his having been at the pains to prosecute his researches by visiting not only the chief sites of the British herring-fisheries, but also the shores of the Baltic and of the German Ocean, besides residing in Norway, and inspecting the principal fishing districts of that country. He has also visited the principal fishing ports of France from Dieppe to Marseilles.

We willingly acknowledge the value of the information thus acquired, while demurring to the rather too self-complacent tone of the following sentence in his preface:—" The author believes he has satisfactorily solved the hitherto disputed questions as to food, periodical visits, migration, &c.; he has also, for the first time, established the important fact that herrings visit our coasts twice in the year; that, in fact, there is a winter and a summer herring periodically arriving on the different coasts, and already, from this knowledge, additional supplies have been obtained where no previous fishery existed." After perusing this, we are not surprised to find Mr Mitchell disputing the authority of sundry statements made by such well-informed writers as Cuvier, and his editor, Professor Valenciennes, and Messrs Yarrell and M'Culloch. We do not think him right in demanding proof of this statement in the 'Dictionary of Commerce' by the last-named author: "There is perhaps no branch of industry, the importance of which has been so much overrated as that of the herring-fishing. For more than two centuries, company after company has been formed for its prosecution; fishing villages have been built, piers constructed, boards and regulations established, and vast sums expended in bounties, yet the fishery remains in a very feeble and unhealthy state." As to the concluding assertion, Mr M'Culloch, in his notes on Adam Smith's 'Wealth of Nations,' has evidently seen cause to modify it, for there he thus writes:—"The character of British herrings now stands deservedly high, and the fishery is become a source of profit and employment to a considerable number of people." We cannot, therefore, allow that Mr M'Culloch unreasonably depreciates the herring-fishing, which, in the opinion of Mr Mitchell, exceeds in worth "the auriferous deposits of gold-diggings." When a man has a hobby for anything, as the Belgian Consul at Leith has for herring, we like him none the worse for riding it boldly and taking the field against all comers. But when he demands proof of the assertion that the herring-fishing has been overrated, it appears to us that his zeal has blinded him to what he himself declares—"in truth, the herring-fishing has become prosperous in spite of every obstacle thrown in its way by the erroneous Government exactions and prohibitions." That this is true we have superabundant demonstration in his tediously lengthy "chronological history of the herring-fishery." And as Mr M'Culloch makes specific mention of "bounties," and cannot be supposed ignorant of the Government follies acknowledged by Mr Mitchell, we, bearing this in mind, hold him justified in maintaining that the herring-fishing has been overrated. The fact is, that partly from envy of the Dutch, partly from erroneous notions as to the power of Government to supersede individual enterprise, and nurse a neglected industry into sudden vigour, the most absurd expedients have been resorted to in order to stimulate the herring-fishery; and foolish expectations of turning the ocean into an El Dorado have repeatedly turned out as futile as the great South Sea bubble.

The encouragement of the fishery by means of a bounty has been the chief obstacle to its prosecution; and this folly was persisted in long after Adam Smith had shown that Parliament had been grossly imposed on in regard to the benefits said to result from the system. During the herring mania of last century, political philosophy was disregarded, while demonstrating that though the Dutch, at a distance from the chief resorts of the herring, were right to employ "busses" (covered vessels of from twenty to eighty tons), boats, capable of following the migrations of the fish into our estuaries, bays, and arms of the sea, were the most suitable for the British fishermen. Never was there a more obvious *non-sequitur* than was involved in the popular argument: "Holland is enriched by herring; Amsterdam is admitted by the Dutch to be founded on herring-bones; the Dutch employ busses, therefore, if we do so likewise, we shall more than rival their prosperity. Let Parliament, then, grant a liberal bounty upon busses." To this disjointed reasoning Parliament unfortunately gave ear. Busses swarmed along the British coasts; Scotland alone, in the year 1776, sent forth no less than 294 of them. But there was no increase in the number of the captured herring, because, as Adam Smith sagaciously observed, "the bounty is proportioned to the burden of the ship, not to her diligence or success in the fishery; and it has been too common, I fear, for vessels to fit out for the sole purpose of catching, not the fish, but the bounty. In the year 1759, when the bounty was at L.2, 10s. the ton, the whole buss-fishery of Scotland brought in only four barrels of sea-sticks" (herrings caught and cured at sea).

When we add that every barrel of buss-caught herrings cured with Scotch salt, when exported, cost the Government 17s. 11¾d.; and that every barrel cured with foreign salt, when exported, cost Government £1, 7s. 5¾d.; we think our readers must allow that this was a clear case of gold being bought too dear, and that it is no marvel that a political economist like Mr M'Culloch should denounce the over-appreciation of herring so caught. The collateral advantage of rearing a race of hardy seamen to man the royal fleet at the demand of his Majesty, could never justify such lavish expenditure, which, moreover, it was foretold, must infallibly extinguish the industry it was intended to foster. The bounty system has ceased, herrings are caught by the million, and her Majesty's fleet is at this day manned by as fine a set of sailors as ever ate sea-biscuit.

We differ, therefore, from Mr Mitchell when observing, "As to the expenditure of the public money on bounties and premiums, it may be seen that the public money hitherto expended has been of a comparatively small amount. It was to enable our own busses to compete with the foreign busses on equal terms." With the herring at their door our fishermen had a great advantage over the Dutch, who had to fit out large vessels to enable them to fish on our distant coasts; and the attempt to stimulate home industry by means of bounties was far from remunerative, when we find it proved that one year, with a bounty of £2, 10s. per ton, the buss fishery of Scotland yielded " only four barrels of sea-sticks."

Mr Mitchell is surprised to find the author of the article "Ichthyology," in the last edition of the 'Encyclopædia Britannica,' adopting Pennant's erroneous theory as to " herrings coming from the icy ocean " to this extent, that he quotes it, and says : " In truth, we are not furnished with sufficient data to decide the question ; but, in the mean time, we do not feel inclined entirely to reject the generally received opinion, that the herrings migrate from north to south in summer and autumn ; " and he then proceeds to describe the " vast troops " which Pennant so fabulously mentions. He says, " The shoals are generally preceded, sometimes for days, by one or two males; a very difficult fact to ascertain." Very true ; but from the ambiguous way in which Mr Mitchell expresses himself, we are in doubt whether this fact rests on the authority of the writer in the Encyclopædia or on that of Pennant. It is very silly to make such an assertion, as no human being can be sure that two big herrings are " acting as guides " of the shoals which they seem to precede ; but having Pennant's statement lying before us, we find in it no reason for fathering this folly upon him. It seems to have been added to it by some lover of the marvellous. Moreover, we find reason for questioning the truth of the general belief that this estimable author was the first to propagate what is now deemed a mistake as to the source whence annually come forth those herring-shoals which make the waves of the ocean resplendent with their glittering hues, and supply a never-failing harvest for the sustenance of man, as well as of numerous orders of birds and fishes. Turning to 'Collection Complète des Œuvres de Charles Bonnet,' the eminent Genevese philosopher and naturalist, we discover the apparent source of the now reputed apocryphal migration of the herring from the regions of "thick-ribbed ice."

Pennant, an accomplished and benevolent Englishman, did good service to Scotland by making its resources known to his countrymen south of the Tweed, and by helping to cultivate a good understanding betwixt the people of North and South Britain. He paid particular attention to what was needed for the development of the Scottish fisheries. His journey to Scotland and his voyage to the Hebrides, begun in 1772, were completed in 1775. Attention to this date enables us to do justice to a worthy man, and to prove that, as he has not been the last, assuredly he was not the first to give currency to the assertion regarding the arctic-pole-haunting habit of the herring. Here is his statement: after representing them as coming from their great winter rendezvous within the arctic circle, he proceeds thus:—

"They begin to appear off the Shetland Isles in April and May. These are only forerunners of the grand shoal which comes in June; and their appearance is marked by certain signs, by the numbers of birds, such as gannets and others, which follow to prey on them. But when the main body approaches, its breadth and depth are such as to alter the very appearance of the ocean. It is divided into distinct columns of five or six miles in length and three or four in breadth, and they drive the water before them with a kind of rippling; sometimes they sink for the space of ten or fifteen minutes, then rise again to the surface, and in bright weather reflect a variety of splendid colours. The first check this army meets in its march southward is from the Shetland Isles, which divide it into two parts. One wing takes to the east, the other to the western shores of Great Britain, and fill every bay and creek with their numbers. Others pass on to Yarmouth, the great and ancient mart of herrings; they then pass through the British Channel, and after that in a manner disappear. Those which take to the west after offering themselves to the Hebrides, where the great stationary fishing is, proceed towards the north of Ireland, where they meet with a second interruption, and are obliged to make a second division. The one takes to the western side, and is scarce perceived, being soon lost in the immensity of the Atlantic; but the other, which passes into the Irish Sea, rejoices and feeds the inhabitants of most of the coasts that border on it."

This is so minute as to convey the idea that, to be accurate, the writer of it must have been amphibious, and that he spent his days either subaqueously or herring-hunting by means of the water-telescope—a valuable instrument used along the coasts of Norway, and deserving the attention of British fishermen. The fact, however, is, that Pennant derives his notions of the pole-haunting herring from Bonnet, whose 'Contemplation de la Nature' first appeared in 1764, and was speedily translated into most of the languages of Europe. As Pennant in 1765 visited France, *Switzerland*, Holland, and part of Germany, and was intimate with Buffon, Haller, and the Gesners, we think it more than probable that at Geneva he made the acquaintance of the ingenious inquirer Bonnet, or at least was familiar with his writings. In the work of Bonnet we trace the origin of the for-long-generally-received belief regarding the migration of the herring. Our translation of a portion of it makes this apparent.

"Herrings migrate in great shoals from the north pole to the coasts of England and Holland. These migrations appear to be caused by the whales and other large fish of the arctic seas, which pursue the herrings. These sea-monsters swallow at once whole tons of them. They often pursue their prey as far as the coasts of England and Holland. The herrings seem to be a manna prepared by Providence for nourishing a great number of fish and sea-birds. In order that the species may be preserved, they must have the power of withdrawing from the pursuit of their enemies.

"The herrings arrive on the coasts of England and Scotland about the beginning of June. These numerous legions then separate in several divisions; some direct their course to the east, and others to the west. After having moved about for a while, the different shoals separate again and diffuse themselves among the shores of the seas of Britain and Germany, reunite afterwards, and finally disappear at the end of several months. It is not merely to withdraw from the pursuit of the large fishes which haunt the seas of the north that the herrings advance in such shoals towards the seas of England and Germany, it is also for the purpose of gathering there the abundant food which nature has prepared for them on these shores, which swarm with worms and small fish, on which the herrings greedily feed, and on which they grow very fat. They at the same time spawn there; and it may be that they forsake the seas of the north because they no longer find in them food sufficient for their subsistence. We may assuredly conjecture that in this respect herrings and other migratory fish resemble birds of passage. The same instinct and the same necessities may determine the migrations of both; and these migrations are the means employed by the wisdom of nature for the preservation of these species so individually numerous."

Now that we have traced the long-prevalent notion of the herrings' annual procession from the north pole, and succeeded, we think, in fathering it not on Pennant but on Bonnet, let us examine wherein the Genevese naturalist differs from the conclusions of more recent inquirers.

Few will question his hypothesis, that the migrations of the herring are regulated by the same causes which occasion the migrations of birds—viz., the desire of procuring food for themselves, and convenient resting-places for the propagation of their species. But as to the tons of herring swallowed at a single gulp of a whale, there are serious difficulties in the way of our gulping that. In the first place, the true whale (*Balæna mysticetus*) of the polar sea is known to feed on the swarms of various species of *mollusca*, especially *medusæ*, which alone are suitable for its singularly limited powers of deglutition, which, it is asserted by arctic voyagers, could not embrace any object larger than an egg. However great the havoc which other species of whales commit on the herring-shoals, if *Clupea harengus*, the herring of our coast, wish to lead a peaceful life, he should remain in the circumpolar waters, and have no dread of the great Greenland whale. As this huge creature does not molest the herring, the southward migration of the latter cannot be owing to his dread of being swallowed wholesale—by tons at a time—as Bonnet fancies. If bent on such a gratification, there is reason for doubting where he could find them. Herrings are nearly unknown within the polar seas, and have scarcely been observed by the arctic voyagers, and they are not taken by the Greenlanders. A small variety of the herring is sometimes found, and is noticed by Sir John Franklin.

As already indicated in our extract from his preface, we are surprised to find Mr Mitchell writing as if he had made important discoveries. Having read his work with care, we acknowledge its value as a repository of facts; but we cannot admit his claims to be regarded as an original observer, whose researches solve the hitherto disputed questions in the natural history of the herring. So far back as the year 1814, we find the writer of the article "Fishes," in the 'Edinburgh Encyclopædia,' thus combating Mr Pennant: "For what purpose they should have received an instinct to retire to the polar seas, is to us incomprehensible. The salmon, the shad, the smelt, are never found at sea, yet it is never said that they depart to any great distance from our shores. The most reasonable conjecture we can form is, that the herring, like our other migratory fishes, takes to the deepest parts of the ocean." "In migrating from the deep seas to our shores, the herring seems to be prompted by a similar instinct to that of the shad and salmon of casting its spawn in its native waters; however, they are more desultory in their movements than either of these fishes."

"From their first arrival in July they keep along both the east and west coasts of Scotland; and in October, after many erratic movements, they fix their residence where they mean to spawn. In these places they continue till the end of February (sometimes, but rarely, longer), and constitute what we call our winter fishing. In the Frith of Forth, for these several years past, this has been a very productive fishery; and during the present winter, 1814-15, the numbers of herrings there taken and brought to Edinburgh markets have yielded a most abundant supply of nutritious food for the poorer class of the inhabitants of the city and its neighbourhood."

In like manner the late Dr Fleming, author of "Ichthyology" in the same Encyclopædia, writes: "It is now clearly established that the herrings, like all the other fishes that reside in deep water, approach the neighbouring shores when they are ready to spawn, and return to their favourite haunts when the process of reproduction is finished. The food of the herring consists of the smaller crustacea, and of young fishes, even of their own species." Similar statements are made by Sir Humphry Davy in 'Salmonia,' and by Mr Yarrell in his 'History of British Fishes.' The latter author states that "three species of herring are said to visit the Baltic, and there are three seasons of roe and spawning. The stromling, or small spring herring, spawns when the ice begins to melt; then a large summer herring; and, lastly, towards the end of September, the autumn herring makes its appearance and deposits its spawn." The same distinguished naturalist has discovered what he thinks a second species of British herring: it is found heavy with roe at the end of January, and does not spawn till the middle of February.

So far from Mr Mitchell being the first to "solve the

hitherto disputed questions as to food, periodical visits, migration," &c., of the herring, the preceding references prove that all that he says on these matters has been said habitually, for the last thirty years at least, by all naturalists of repute among us. And as to his being "the first to establish the important fact that herrings visit our coasts twice in the year—that, in fact, there is a winter and a summer herring periodically arriving on the different coasts"—we are certainly indebted to him for drawing attention to this habit of the fish; but we must be excused for declining to acknowledge that he was the first to observe it.

The capriciousness with which herrings change their haunts has given rise to many strange speculations as to the cause of a characteristic so annoying to those dependent on the regularity of its migrations. By not making prudent allowance for this peculiarity, and by constructing establishments as if for a fixed fishery, serious losses have been frequently incurred. From the Friths of Forth and Tay not only herring, but also haddock, took their departure about the year 1788, and did not reappear for nine or ten years.

In one of the earliest of Hugh Miller's productions descriptive of the Moray Frith herring-fishery, there is a notice of this peculiarity. From 1690 to 1709 there was an extensive fishery at Cromarty. Shortly after the Union (1707) an immense shoal ran themselves on shore in a little bay to the east of the town. The beach was covered with them to the depth of several feet, and casks failing the packers, the residue was carted away for manure by the neighbouring farmers. They disappeared in a single night, and were not seen again for more than fifty years! And so with very many other places frequented by the herrings. Of course they have good reasons for what we call their capriciousness. But if herrings seem to be fickle, fishermen are not seldom stupid and lazy. They groan over the departure of the fish instead of carefully watching their movements. There is reason for believing that a little enterprise would often be rewarded by the discovery that the fish had moved to no great distance, and that their change of habitat might be traced by observing that luminosity of the sea caused by medusæ and other marine animals on which the herrings feed.

As already mentioned, the speculations in supposed explanation of the wayward movements of the herring are many. In illustration of the propensity of certain minds to join together, in the relation of cause and effect, things which are only accidentally connected, so that coincidences are mistaken for consequences, it may be mentioned that in the Hebrides the manufacture of kelp is popularly regarded as the reason why the herring deserted the Long Island; and that the lighting of fires, the ringing of bells, and the firing of cannon, are all associated with the vagaries of the herring. Will people never learn to be decently modest, and not afraid to say, "Such is the fact, but why it should be so we cannot tell"? When not ashamed to confess ignorance, we shall no longer read that herrings forsook the coast of Sweden on account of the din of the bombardment of Copenhagen; or that "Inverness, where sumtym was grit plenti of tak of herring, howbeit they be now evanist for offens that is maid against some sanct;"* or that the people of St Monance, on the Fifeshire coast, were wont to tie up the parish church-bell during the herring season. If the theory at Inverness as to wrath of "sum sanct" be admissible as an explanation why herrings did not appear as expected, we must say the men at St Monance, by tying up the bell which summoned them to do him honour, exhibited a love of herrings which must have been offensive to his holiness. The theory of steamboats having something to do with the herring deserting their haunts, is falsified by the fact of herring preferring Loch Fyne since steamships have sailed to and from Inverary.

* Bellenden's 'Bœce's Cosmographie of Albion,' xxiii.

Instead of wasting time assigning fanciful reasons for what is known to be a fact in the natural history of the fish, it will be more to the purpose that we should consider whether, as in the case of salmon, the migratory instincts of the herring can be so guided as to make this fish inhabit waters which it has not heretofore frequented, or return to those which it has for a time deserted.

Its habits, there is reason for believing, are such as to encourage experiments in pisciculture such as have thrown so much light on the habits of the salmon. If rivers, heretofore destitute of this most valued of the fresh-water fishes, are no longer salmonless, we are encouraged to hope that the coast-haunting herring, if introduced in any favourable locality, may be found as regular in its periodic movements to and from the ocean-deeps as is the salmon in its migration from the sea to the river. We are glad to learn that "Lacépède says, that in North America the spawn of the herrings has been carried by the inhabitants and deposited at the mouth of a river which had never been frequented by that fish, and to which place the individual fishes from these spawn acquired a habit, and returned each year, bringing with them probably a great many other individuals of the same species." "It might," observes Mr Mitchell, "perhaps add to our knowledge of the natural history of this animal, if some of the proprietors of sea-water fish-ponds were to make experiments in the same way, or even by transporting the herring alive. The said author also states that in Sweden they have been transported alive to waters where they were wanted."

This is a statement equally interesting and important, and if Lacépède gives any details, it is to be regretted that they have not been fully given by Mr Mitchell, whose account of the mode in which the herring spawns, though apparently accurate enough, does not appear to throw upon it all the light that is desired, seeing that, so lately as 1862, the Commissioners of the Fishery Board employed men accustomed to use the diving apparatus to examine the localities where the herrings were supposed to deposit their spawn, and to bring up portions of it for examination. Various places in the Frith of Forth were examined without success, but at last, east of the Isle of May, a considerable quantity was found in twenty fathoms water, adhering to coarse shelly sand, the deposit being about three-fourths of an inch thick, and attached to a cake of the rough shells and sand. Though the divers saw in the vicinity of this spawn fishes like herring moving about, they did not see them so distinctly as to enable them positively to assert that they were herrings. The ova thus obtained produced little fish, exhibited at a meeting of the Royal Physical Society, and some of them lived in an aquarium about four weeks. Attempts also were made to breed them by placing the ova in boxes or cans in the sea, but unfortunately these were destroyed. There is thus only a strong probability that the vivified ova were those of the herring. The defects in this experiment are not thoroughly removed by Mr Mitchell's account of the manner in which the herring spawns. After describing the deposition of the ova on stones, rocks, and sea-weeds, to which they firmly adhere, he thus proceeds:—" The eyes are first observable —at least a small black speck is first seen in the egg. Then the head appears; and in fourteen days, or *perhaps three weeks*, the young are seen in great abundance near the shore, of a very small size. In six or seven weeks more they are observed to be about three inches in length, and move about in large shoals in winter and spring on the various coasts, and in the rivers and bays generally resorted to by the herring-shoals; and *it is likely* that they attain to full size and maturity in *about* eighteen months."

When we find, in the most recent work on the herring, such indefinite phrases as those we have italicised;

when the period at which the fish reaches maturity is only guessed at by reduplicated peradventures, we are not "perhaps," but positively of opinion that our knowledge of the natural history of herrings requires enlargement; and we know of no mode of attaining this desideratum so effectually as by so placing the procreant fish that they shall constantly be watched by intelligent observers. Until its reproductive processes were studied in rearing-ponds and boxes, how little did we know about the salmon! what nonsense was written about it, too, hurrying to and fro between the arctic seas and those of Great Britain! We are persuaded that similar experiments on the herring may be made with little difficulty; and this department of our fisheries being one in which Government already interferes, it is much to be desired that it should forthwith institute a piscicultural establishment, charged with the duty of specially investigating the natural history of the herring, alike in its natural habitats in the ocean, and in others selected for the convenience of scientific observation. It is all very well in my Lords of the Privy Council for Trade to send circulars to the fishing-stations, soliciting replies to queries anent herrings. We wish they may get them! The Dutch Government acts more intelligently, and with more patriotism, when authorising the Royal Meteorological Society of Holland to obtain information regarding the natural history of the herring; and so important is the information given in a work published by authority of that Government, that the British Board of Trade has caused it to be translated into English. The British and the Dutch have had many a tough battle about fisheries; and now that we are such good friends as to interchange books about fish instead of hard blows, we hope there may be a reciprocity in the interchange of information, and that we shall not be humbled by the Dutch continuing our superiors in ichthyological investigations, to engage in which we have such special inducements.

The fact is, though the strife be peaceful, the Dutch are at this moment contending with us, seeking to regain the prestige of their herrings, gradually diminished in consequence of the improved quality of the Scotch cured herrings. The number of busses employed by the Dutch has suffered very serious diminution, and the supply of fish has likewise been lessened in even a greater proportion. The herring-fishing of this country, on the contrary, and especially of Scotland, has made signal progress, and constitutes a branch of national industry deserving to be developed by all the appliances which can be brought to bear upon it by the combination of enterprise and capital guided by intelligence. Of the 91,139 people directly employed in the Scottish fisheries, 39,266 are fishermen; but, observes Mr Mitchell, —" If we add those employed indirectly by the money derived from the fishery—namely, the boatbuilders, sailmakers, ropemakers, mastmakers, saltmakers, grocers, carters, porters, shipowners, sailors, and other trades—the number will appear incredible to those who have not an opportunity of closely observing the incalculable benefits accruing to the nation from the prosperous state of such a fishery. Here we see employment to the industrious classes, while they are adding an abundant supply of cheap and wholesome food for the numerous population of the British Islands, when other animal food is becoming so scarce and expensive.

" The great increase of this fishery has tended in no small degree to increase the wealth and the number of the population of the Scottish coasts; and the annual addition of the value of the herrings must have a great and beneficial influence on the prosperity, not only of Scotland, but of the British Islands. The addition of nearly one million sterling every year to our national wealth must be extremely gratifying to every patriotic mind."

In the reign of Henry VIII. one of the statutes thus recognises the importance of the fishermen of Kent and

Sussex: "The which said mariners were put in daily experience and knowledge of the coasts of the sea, as well within this realm as in other parts beyond the sea, by the which practice it was great strength to this realm, by bringing up and increasing of mariners, whensoever the King's grace had great need of them." On this ground, when abstinence from flesh ceased to be considered a religious duty, it was attempted to keep up the consumption of fish by statutes passed at various times. The Act for the abstinence from flesh (passed in 1548) imposes penalties on persons eating flesh on "fish-days," of which in the year there were no less than a hundred and fifty-three.

The real object of all this intermeddling with the diet of the people being "for the maintenance of the navy," it is unpleasant to find Parliament hypocritically preaching up abstinence from butcher-meat as "a means to virtue, and to subdue men's bodies to their soul and spirit" (2 and 3 Edw. VI. c. 19). We admit that fish "was much behoveful and necessary to the common weal of this realm" (1 H. VIII. c. 1), and that it was a scandal that "the natural subjects of this realm are not able to furnish the tenth part of the same with salted fish of their own taking" (39 Eliz. c. 10); but while clearly of opinion that the Crown, as the feudal possessor of the shores of this kingdom, ought to develop the inexhaustible resources of this magnificent domain, we, of course, object to all needless interference with the fisheries, and the modes in which fishermen prosecute their calling. Notwithstanding the recent agitation for putting down the system of branding the prepared herrings, and thus giving a guarantee to purchasers that they are of a certain quality, impartial judges are of opinion that the system ought to be continued. We are disposed to go further, and to maintain that it is an important function of the Fishery Board to establish a coast-guard so watchful that we shall not again hear of such senseless destruction of spawning-fish as occurred near Dunbar in September 1861. A shoal of herring having begun to spawn at a short distance from the harbour, were immediately assailed by the fishermen with such indecent greed, that on Sunday 1st September, and the two following days, the neighbourhood had the appearance of a fair, so great was the bustle in disposing of the most improperly captured herrings. This is monstrous. If a man were to open a shop in Dunbar for the purpose of shaving or selling "sweeties" on *Sunday*, we have no doubt that there would be an outcry. But when lawless herring-slayers profane the Sabbath, they are, if not unblamed, at all events unmolested, which is all they care about! Other folks besides the fishermen of Dunbar are apt to leave their sanctity on shore during the herring season. We once heard a Gaelic minister threaten to keep back from the communion-table all who should not abstain from fishing on the fast-day. When fishing is thus prosecuted "in season and out of season," we cannot be surprised to learn that many of the Highland sea-lochs, answering to the Norwegian fiords, no longer roll abundance to the very door of the now almost starving Highlander, whose unfair and unseasonable fishing has driven the persecuted herring from its haunts. In Norway, where 40,000 men are engaged in catching the herring, every facility which science and ingenuity can suggest is offered to its spawning in the numerous fiords; and the spawn and fry are carefully protected from molestation, in order that the fish may keep upon their coasts, and breed in security. But our Scottish Fishery Board is loudly accused of being afraid to put in operation the law against trawling; and there is reason for questioning the discretion of the Board in availing itself of the power to suspend any of the clauses in the recent Act. In January 1861 the clause prohibiting trawling with small-meshed nets was suspended, for the purpose of permitting the fishermen of the Forth to catch sprats, or garvies as they are called in Scotland; and it is asserted that, on the morning after the removal of this wise restriction, four fishing-boats brought in and sold as garvies above 400,000 herring-fry. Had these fish been spared for a few months, and sold as herring at the rate of two a penny, they might have brought about £880, while as garvies they probably did not sell for more than as many farthings. We are aware that a report by Professor Huxley, Dr Lyon Playfair, and, we think, Mr Caird, urges that the slaughter of sea-fishes by all means invented by human ingenuity is so comparatively trifling, that it is ridiculous to interfere with trawling and garvie-catching. We read it with astonishment, and now protest against it as sanctioning a loose morality and reckless disregard of the future by no means to be encouraged. We are persuaded that it *is* possible to make fatal inroads upon the herring, type of inexhaustible fecundity though it be considered. Able ichthyologists maintain that the sprat or garvie is not the young of the herring, the distinction between them consisting in the position of the fins, and in the belly of the one being serrated, while the other is smooth. We are at a loss what authority to attach to the statement of an anonymous writer in the 'Cornhill Magazine,' who, from experiments he has made, insists that the latter peculiarity leaves the fish as it grows older, and that the sprat or garvie is really the young of the herring. Remembering the long ignorance of even the scientific in regard to the progeny of the salmon, we think it worth while that this statement shall be tested; and having heard Professor Huxley arguing that the archetype of the human frame is that of the monkey, we presume that he, at least, will not be astonished if it be shown that the herring is merely a developed garvie.

There is another point as to which Government influence and encouragement may tell most beneficially upon the fisheries, as well as upon the hardy race of men by whom they are prosecuted with an amount of mortality which certainly should figure in the returns of the Registrar-General as "preventible." Every now and then the voice of wailing is heard along our coasts because of that loss of the crews of fishing-boats which justifies the Scottish poet when he makes the fisherwoman say,

> "Buy my caller herring;
> Though ye may ca' them vulgar fairing,
> Wives and mothers maist despairing,
> Ca' them lives of men."

We allude to the obstinacy of our fishermen in plying their hazardous calling in undecked boats. Caught in a storm, there are on the east coast few harbours into which they can run; and so these boats are driven helplessly out to sea, where they are swamped and lost, after the men have been exhausted in vain efforts to keep them to the wind in order to avoid shipping the broken waves which surround them. Heavily laden, a single wave may swamp them; and when this casualty overtakes a whole fleet, the scene of misery which ensues is indescribably distressing. In 'Life in Normandy' there is a fearfully graphic picture of the agony of the watchers on the beach near a Scottish fishing village straining their eyes to catch a glimpse of what was befalling their nearest and dearest in mortal jeopardy on the raging sea. "Whisht ye," said an old woman to a boy crying aloud by her side; "keep yer greetin till it's wanted; Lord knows but ye may be a faitherless bairn before mornin', and I a bairnless mither." Dismal forebodings sadly verified! Thirty-two boats were lost, and ninety-one poor fellows perished; and this was not all—numbers of the fishermen were ruined by the loss of their nets.

"All this loss of life and property arose from the boats being open; for every boat that was lost either foundered at its nets or was swamped in the sea. Had they been decked, the men could have hung on and taken their nets on board without fear of foundering. A few seas might break on board of them, but with a deck the boat would rise again as the sea rolled past. Give me a lid to the pot; and I wish I could only persuade our countrymen to be of my opinion on this subject, and copy the Frenchman."

This wish is echoed by a Northumbrian fisherman, with whom Mr Campbell has a most amusing colloquy. "A Frenchman," quoth he, "is precious pig-headed, but I suspect our own people have a touch of the same nature too, for they will stick to the open boats. They say the boats are lighter and more handy, if they have to take to the oars, when they have no deck; and, no doubt, they are right in that, for a deck must always weigh something, be it ever so light. But these Frenchmen shove along very well with their oars, and the devil's in it if a Scotchman cannot do as well as a Frenchman. It's prejudice and laziness that make them stick to a plan that risks their lives and properties far more than they need to do."

So long as they are so lazy and ignorant as they generally are, we shall speak to them in vain of the advantage of a small steam-tug being employed to convey with rapidity their boats to and from the spots frequented by the herring. And, we fear, it will be hard to teach them the value of life-buoys, or safety-belts, or persuade them to learn to swim. We have long been aware how few fishermen have acquired the art of swimming; and being at sea in a herring-boat the summer before last, we had a curious talk with the crew on this matter. Being cold, we said to the friend who accompanied us, "Come, let's have a row." Being clergymen, and, in the opinion of the fishermen, landlubbers of course, they were not a little surprised to see us set about rowing like men who knew how to manage an oar. "What are you staring at?" asked one of us. "Oh, we never saw twa ministers rowin' a boat afore!" "Indeed! my father had more sense than yours, I suspect. He was a minister, but he taught me to do three things—ride, row, and swim. Can any of you swim?" "No, sir." "Why, man, if you fell out of your boat in smooth water even, you would be drowned!" "The Lord could save me, sir." "Oh, yes, He *could*, but He won't. Do you think He will work a miracle to save a man so lazy as not to have learned to swim?"

Though there must be great difficulty in either Government or private persons indoctrinating fishermen with ideas foreign to their experience and modes of thought, the effort to enlighten and assist them so far as proper should be perseveringly made. Government may regulate the size and equipment of their boats, and afford them the protection of large vessels and steamers at different points during the busy season, besides providing charts indicating the best fishing-grounds, and the nature of the tides, currents, &c. Above all, it is bound to be liberal in aiding the construction of harbours of refuge, the existence of which will go far to prevent those scenes of disaster which so often fill fishing villages with lamentation, and entail upon the poor-rates the heavy burden of maintaining families bereaved of the support wont to be supplied by the rough hands which laboured hard to "win the bairns' bread."

POPULAR WEATHER PROGNOSTICS.*

"To popular apprehension, the highest or ultimate object of meteorology is to enable us to foretell the weather. Looked upon in this point of view, science can as yet only offer abortive attempts, or such as hold out no promises for the future. In other respects, on the contrary, her advances have been assured, rapid, and brilliant." We shall not presume to speculate whether the illustrious Arago, if still among us, and acquainted with Fitzroy's forecasting of the weather, would have been induced to modify this opinion as to the hopelessness of expecting that scientific meteorology shall enable us to foretell the weather. It is certain, however, that he would have approved of the attempt now being made, by the Scottish Meteorological Society, to determine the worth of that "folk-lore" regarding the weather which has been current from the earliest times. "While," says he, "I am far from regarding proverbs and popular sayings in general as constituting codes of national wisdom, I am, at the same time, disposed to

* 'On the Popular Weather Prognostics of Scotland,' By Arthur Mitchell, M.D., Member of Council of the Scottish Meteorological Society, &c. William Blackwood and Sons, Edinburgh and London: 1863.
'A Prognostication of Right Goode Effect.' By Leonard Digges. London: 1555.

think that physicists have been wrong in treating with contempt those sayings or proverbs which refer to natural phenomena. It would, no doubt, be a great mistake to receive them blindly, but it is no less so to reject them without examination. Guided by those principles, I have sometimes found important truths where others had seen merely groundless and obstinate prejudices."

It was in this spirit of rational deference to the experience of the unlearned that the Marquess of Tweeddale declared to the Meteorological Society that certain shepherds, without scientific apparatus, and simply by watching the aspect of the sky and the atmosphere, had been able to foretell the extraordinary snow-storm of 1861 in sufficient time to save themselves and their flocks from its disastrous effects. "He thought these natural observations ought to attract a much greater amount of attention than they had yet received; and as Dr Arthur Mitchell and himself were at present engaged in collecting such observations for the purpose of being classified and systematised, he hoped that they would receive every possible assistance from their country friends in their inquiries."

Very possibly the determination of the Society to enter upon the investigation of this very curious and interesting department of knowledge may excite more popular notice than has been excited by any of the important inquiries which it has been prosecuting. We hope that it will, and that the result will be an accession of reliable information to the public, and of much-needed funds to the Society. And should this hope be realised, the public and the Society will have equal reason to be grateful to the Marquess of Tweeddale for his valuable suggestion, and for his liberality in offering a prize of twenty guineas for the best scientific explanation of the prognostics collected by Dr Mitchell.

If few of our readers be sufficiently learned to compete for the prize, we hope that many of them will, for their own instruction, and as a contribution to the facts to be inquired into, set about testing the value of the prognostics relied on in their neighbourhood; and if they attempt to explain them on scientific principles, we venture to assure them that the result will be a very remarkable demonstration of the amount of their own ignorance! The *rationale* of these phenomena must be sought for in intimate acquaintance with "*rerum naturâ*," as Lucretius phrases it,—in other words, in knowledge of the laws which bind together all things animate and inanimate; sun and moon, rainbow and aurora borealis, falling stars and thunder; the aspects of mountains and of the sky; the varying sounds of the ocean, rivers, and cascades; subterranean noises, and the issue of gases and foul air from the crevices of mines; the movements of various plants; the proceedings of birds, fishes, and quadrupeds; the sensations of human beings; and a host of unclassified prognostics. The man who thinks himself qualified to inform us intelligently how all creation here below is in sympathy with that all-pervading something called the weather, must be a graduate of a university which communicates to those attending it a great deal more knowledge than we got at Edinburgh College, or have managed to pick up since in the great school of daily life. Dr Mitchell's collection of weather-signs has interested us much, and, to tell the truth, has humbled us greatly.

'Guesses after Truth, by Two Brothers,' is the modest title of a very delightful modern book; and if the meteorological guesses of the competitors for the Tweeddale prize find as many admirers, it will be worthily bestowed. We confess that in so wide a field we expect many explanations of meteorological phenomena which will only deserve to be called guesses; but that is no reason why these phenomena should not be classified and seriously examined: the guesses of the wise of one generation are often, to the next, demonstrated truths of great importance. And these popular prognostications of the weather, whether true or false, are important, as Dr Mitchell truly observes:—" Faith in such signs determines in no small degree the *actions* of shepherds, farmers, seamen, and others, by whom they are trusted in such a manner as to lead either to gain or loss. It is this consideration which gives to the study of them a practical value. They either mislead and cause loss of time and property, or they are useful and ready guides to be consulted and obeyed with profit. Their actual influence on conduct for good or evil makes it clearly desirable that their trustworthiness should be carefully tested."

It can hardly be doubted that there are natural indications which, when long and accurately observed, give certain premonition of approaching atmospheric changes. No classes of men are so much interested in this kind of knowledge as farmers, sailors, and shepherds, and none have more ample means of observation. Whether farmers are generally as observant of weather phenomena as their interest should prompt them to be, may be questioned, considering how few of them are members of the Meteorological Society. Sailors, for their own safety—shepherds, for that of their charge—are notoriously weather-wise. "I remember," writes a friend of ours, "of being in company with the celebrated Sir Sidney Smith, when he visited Perthshire, many years ago. It was in autumn, and the grain was ripe in the fields, but the weather was so unfavourable that the labours of the harvest were at a stand. I observed to him that nobody understood the weather so well as sailors and shepherds, and I hoped he could give us some reason to expect a change. 'Oh!' said he, with the frankness characteristic of himself and his profession, 'I can do nothing among your mountains; but set me afloat in a known sea, with a barometer before me, and I will give you a rough guess of what you may expect.' And he proceeded to say, that whenever he went to a new station, the first thing he did was to call around him all the oldest fishermen, and mark down all their signs and observations of the weather; and he added that he never found them mistaken."

Weather prognostics such as guide the movements of fishermen, sailors, and shepherds, are not to be laughed at because trusted in by those making no pretension to science. Let the scientific consult their instruments, and, if classically disposed, learn by heart the celebrated Addison's Latin poem, entitled 'Barometri Descriptio;' but let them not despise the *empirical* knowledge of the unlearned, handed down from the days of old, and trusted in because founded on experience.

As we declared, in our article on the eating of funguses, that we would, in a matter affecting the bodily senses, follow the opinion of a savage rather than of the Pope or the President of the Royal Society—so say we now of weather prognostics, that, being matters of fact such as any man living much in the open air can observe, the recorded observations of the illiterate are worthy of careful attention. They relate to matters in which they are daily interested, and, as such, are level to their capacity; and it is mainly to such observers that we are indebted for those prognostics in the explanation of which philosophers have as yet said so little that is satisfactory.

It is to the credit of such observers that our scientific meteorologists seem nowadays seriously disposed to test the value of those popular maxims which appear alike in the writings of ancient philosophers, and in those curious repositories of weather-wisdom—almanacs—which a recent French writer asserts are perhaps the oldest books in the world, except the Scriptures. Aristotle in his book 'De Meteoris,' appears to have been the first of the Grecian philosophers to collect and systematise the various prognostications of the weather; and his speculations on meteorology, though often unsatisfactory, are sometimes—as in his remarks on *dew* —remarkable for their approximation to the discoveries of modern science.

His pupil, Theophrastus, wrote in a more popular style, in a work of four general divisions—viz., the prognostications of rain, of wind, of storms, and of fair weather. Socrates, too, seems to have been something of a meteorologist. The ribald poet Aristophanes held him up to ridicule in 'The Clouds;' introducing him in a basket drawn across the top of the stage, for the purpose, he was made to tell the scoffing mob, of making observations on the weather. The poets, both of Greece and Rome, also treated of the universally interesting subject of weather. Aratus, once the popular versifier of 'Phenomena' and 'Prognostics,' is now chiefly known to us because quoted by St Paul in his speech to the men of Athens, and because of his singular fortune in having had two such illustrious translators as Cicero and Cæsar Germanicus. Facts and fables, follies and superstitions, are so mingled in the meteorological portions of the poetry of Lucretius, that they do not merit serious consideration. No reader of the 'Georgics' needs to be reminded that far higher heed should be paid to the weather-wisdom of Virgil. Believing him to have been an accurate observer, it would be interesting to compare the prognostics which in his day indicated change of weather, with those which are still relied on by the people of Italy; and should some Italian meteorologist be induced to make the comparison, we should anticipate new confirmation of the persistency of the popular faith in natural indications of coming changes in the condition of the atmosphere.

Passing from the philosophy and poetry of the ancients to the kalendars and almanacs of times nearer our own, we come upon a singular repository of strange things about men and animals and the weather. As a contrast to a quarterly report of the Meteorological Society, we think our readers will not be displeased to be furnished with a few specimens of popular meteorology, as set forth in some of the old almanacs of England and France.

We begin with 'A Prognostication of Right Goode

Effect, by Leonard Digges; imprynted at London, within the Black Fryars, by Thomas Gemini, 1555.'

It commences with "many pleasant and chosen rules for ever to judge of alteration of the weather." First, according to the day on which the moon changes: if on Sunday, dry; if on Tuesday, windy; if on Wednesday, wonderful; if on Thursday, fair and clear; if on Friday, mixed weather; if on Saturday, moist weather.

We are also told that the planets influence the weather; and lastly, that the day of the week on which New Year's Day falls determines the general character of the ensuing year. Thus, if on Friday, we shall find "the somer scante pleasant; harvest indifferent; little store of fruit, wine, and honey; corn deare; many bleare eyes; youth shall die; plenty of thunders and tempests; with a soden deathe of cattel."

New Year's Day on Saturday prognosticates "a mean winter; somer very hot; a late harvest; good chepe garden herbes; plenty of hempe, flax, and honey."

As might have been anticipated, the stars, according to Digges, are very potential. The conjunction of the sun and moon indicates "a very unhappy day for all matters; therefore neither plante, build, sow, nor journey." When Venus is in conjunction with the moon, "then is the time to sow, to marry, to follow all manner of pleasant pastimes, and not unmeet to hire servauntes *or to let bloode.*"

In our rather horrifying article on "Hirudiculture" we allude to the incomprehensible tendency of people in the country to part with their own blood periodically. We were recently shocked to hear of a farmer near us bleeding all his young cattle, for no better reason than because it was "gude i' the springtime."

To return from this digression to the weather, we expect the special thanks of agricultural readers for next introducing to their notice a very singular production: 'An Everlasting Prognostication of the State and Condition of every Yeare, by the onely Kalender of Januarie, written by that auntient learned Leopoldus Austriacus, and others, for the commodity of the wise husbandman.' As 1864 began on a Friday, the vaticination of the learned Leopoldus was unusually interesting. Here it is:—"If the first daye of Januarie happen on Fridaye, then shall the winter be very cold and dry, the springe boysterous and wette, the summer temperate, the harvest more wette than drye, so that blear dews and other diseases, with the filthinesse of matter, and running in the eyes, is to be feared; and the *pin* and *webbe* is likewise to be doubted to happen that year. And young children shall then dye, and a likelihood that young women shall be lured into love through the flatterye and great persuasion of men. Also plenty of fruites is then promised, though much haile fall that year."

Our far-seeing, learned man, is also kind enough to give us a peep into the coming year, by teaching us to interpret the state of New Year's Night. "If it be calm and cleare, without winde and raine, then doth the same promise a prosperous year following; and if, the same night, the wind happen to blow out of the east, then doth the same signifie the dearth of cattle; and if, the same night, the wind happen to blow out of the west, then a likelihood of the death of kings or princes to ensue that yeare. If, in the same night, it blow out of the south, it signifieth the death of many persons that yeare; and if it blow out of the north, a small yielde of all the fruits of the earth that yeare."

In Dr Dee's 'Almanac newly set forthe,' in 1571, we have prognostics of the weather as indicated by natural objects,—the redness of the moon foretelling wind— the moaning of the sea, storms—the early flight of swallows, a severe winter. All these, and the inevitable "Thirty days hath September," &c., we find just as in an almanac of this year.

About the end of Elizabeth's reign, almanacs appear

to have become a necessity for all classes. "John Wodehouse, philomath," is not so wise as a man should be with such a title. He tells us that cabbages are to be sown in the wane of the moon, and radishes at the increase; that gillyflowers are to be planted "in an old moon," and parsley sown at the full.

Turning to France, we find the almanac the same strange mixture of the useful with the absurd. M. Nisard, in his work on the colporteur literature of France, gives an interesting account of one of the oldest, which is in vogue to this day in the remoter districts of the French empire. 'The Shepherds' Kalendar'* is profusely adorned with woodcuts alarmingly contemptuous of drawing and perspective. The frontispiece gives us the shepherd's portrait, his bagpipes under his arm, while the shepherds, each with bagpipe or flageolet beside him, are sitting round in various attitudes of attention and wonder, with a dog in the foreground, apparently as much surprised as any of them. The shepherd, prefacing his discourse with the remark that "shepherds who lie in the fields at night see many signs," now commences a marvellous meteorological lecture, in which the sun, moon, and planets play a very subordinate part compared with comets, flaming stars, and fiery dragons. This lecture is profusely illustrated with woodcuts. A comet, very hairy and very fiery, with a tail resembling a broomstick; stars with bats' wings, about to pounce on the astonished shepherds' heads; and an unmistakable dragon, with staring eyes and a most voluminous tail, breathing out volumes of smoke. The shepherd does not say much about them, probably deeming their portraits quite sufficient; he tells us, however, that all manner of mischief is to be expected from their appearance.†

* 'Le Grand Compost, et Calendrier des Bergers, composé par le Berger de la Grande Montagne; fort utile et profitable à gens de tous états.'

† Those desirous of knowing more about such productions should

After perusing such vagaries, one cannot help wishing that there may be truth in Coleridge's assertion, "In the *imagination* of man exist the seeds of all moral and scientific improvement; chemistry was first alchemy, and out of astrology sprang astronomy." In that case meteorology, having for long been the product of imagination, is destined some day to rank among the exact sciences, and confer upon mankind signal benefits.

Finding that almanacs, British and foreign, are to a great extend the chosen vehicles through which the weather-wise communicate with the public, we have been induced to extend into Germany our research after prognostics. We have been rewarded by stumbling upon the 'Göttingen Pocket Almanac for 1779,' in which we find a collection of the most authentic observations of recent writers on "the pre-sensations" which animals have of the weather. Being too ignorant and too indolent to compete for the Tweeddale prize, we benevolently help the essayist by making him acquainted with the observations of Dr F. A. A. Meyer of Göttingen.

He classifies these pre-sensations under three heads—the pre-sensations which animals have (1) of fair or dry weather, (2) of rainy weather, (3) of stormy weather.

As to the pre-sensations which animals have of fine or dry weather, here is the theory of the Göttingen *savant*: Clear dry weather generally follows after wet weather, when the atmosphere has been freed from the vapours collected in it by their falling to the earth in rain. Clouds as well as rain are the means by which the air frees itself from the electric vapours that are continually arising; and if these, again, fall down, it appears very natural that animals which live chiefly in the open air should express, by various movements, the ease with which they breathe and perform all the vital

---

read " Kalendars and Old Almanacs," in the 'British Quarterly Review,' October 1858. It is very amusing, and we are indebted to it for several of the statements in this article.

functions. This pre-sensation is highly useful to bats as well as insects; their wings not being protected against moisture by any oily matter, rain would render them heavier, and unfit for flying.

Dr Meyer supposes that, on the approach of dry weather, larks and swallows fly high because the upper regions of the atmosphere are freer from vapours, and because the insects on which they feed then probably take a higher flight. He also alludes to the weather-fish (*Cobitis fosarlis*) kept in Germany for the purpose of foretelling weather; because, when the weather is fine, it continues quiet, but is restless before rain or storm. As Dr Mitchell does not enumerate fish among the animals whose proceedings prognosticate rain or other atmospheric change, we suggest this as a topic for investigation. Every angler knows how often his sport is marred by a change of wind, by impending rain, and specially by the approach of thunder. In order that they may be always under observation, fish in ponds, or in vessels in the house, should be attentively watched. If the quiescence of a leech in a bottle denote fine weather, and if its mounting to the top of the water and clinging to the sides of the bottle betoken rain— as, from experience, we think to be the case—we should expect to gather weather-wisdom from the pre-sensations of fish in confinement. Speaking of leeches, we remember of being struck with a curious application of the motions of a leech. An ingenious Frenchman, in the great Exhibition of 1851, exhibited "an animal barometer" in the shape of an apparatus so arranged that the movements of a leech were indicated by the ringing of a bell.

As to the pre-sensation which animals have of rainy weather, Meyer supposes that this may be explained by the increasing weight of the atmosphere, by their manner of living, and by the want of moisture necessary to their existence. People who have wounds or old ulcers experience contraction and heat in the parts affected;

why, then, should not the skins of animals be similarly influenced? This, in Meyer's opinion, explains why horses and asses rub themselves, shake their heads, and snuff the air by turning up their noses; why asses bray much and jump about; why cattle scrape up the earth and stamp with their feet; and why swine, though not hungry, eat greedily, and dig up the earth a great deal with their snouts. Our learned instructor from Göttingen may think so; we have our doubts, and beg to enter a special caveat against the libel on pigs, —" Eat greedily, *though not hungry.*" We believe they have more sense. They have good appetites, let us be thankful, otherwise we should not have such good hams. If Göttingen swine behave like gluttons on the approach of rain, we can only say that our magnificent pig has no such evil custom, and eats his meals, foul day or fair day, with an equanimity which makes us envious.

As to the pre-sensation of rain inferred from the increased biting of fleas, of that also we have no personal experience. We admire the modesty which prompts the declaration, "This we cannot explain, as the natural history of this and other similar insects is as yet too obscure." This modesty makes us benevolently hope that rain does not fall in excess at Göttingen, and that the liveliness of fleas is consequently not discomposing to the natives. As to the pre-sensation which animals have of storms, our German Doctor is surely quite astray when maintaining that hitherto this has been observed only amongst the most perfect of *mammalia*—namely, man and the dog; and we doubt not that he was dreaming in the clouds—pipe-begotten—when generalising thus:—"It appears, in general, that the more imperfect animals remark only the approach of dry weather, the more perfect the approach of rain, and the most perfect the approach of storms."

"The dog, on the approach of rainy weather, expresses signs of uneasiness, scratches himself, because the fleas then bite him with more violence, digs up the earth with his feet, runs round, and eats grass; he is accustomed, however, to do the latter when he is very hot, perhaps to cool himself; and, in general, a storm follows soon after." Well, it may be so at Göttingen; and if the professors there report as to the habits of fleas in assailing human beings, we shall listen with deference. These lucubrations as to the doings of fleas with the canine race are open to suspicion, because, though dogs are our good friends, they are unfortunately unable to explain their sensations meteorologically; and as our dog seems to eat grass because he needs physic, we are not prepared to grant that when German dogs are graminivorous the sky will soon growl out thunder. In short, we cannot accept either the facts or the philosophy of Dr Meyer. So far is it from being true that only man and the dog have been observed to have a presensation of storms, that, among the prognostics of quadrupeds, Dr Mitchell has collected the following for investigation:—Goats leaving high and exposed ground, and seeking shelter in a *bield*, or in some recess; old sheep and ewes eating greedily; swine carrying straw in their mouths, and tossing about their bedding; moles raising their hillocks more than usual; hares taking to the open country before snow.

Coming now more especially to some of the prognostics noted by Dr Mitchell for verification, we find that it is believed that the low flight of crows indicates rain, and that if, when flying high, they suddenly dart down and wheel about in circles, wind is expected. We can only say that, if they do, we can suggest no explanation of such proceedings; and that if the crowing of the cock at unusual times prognosticates rain or snow, we are quite at a loss to explain the augury. It is certain, however, that, from the earliest times, ravens and cocks have got credit for being weather-prophets.

The landward flight and flocking of sea-gulls are supposed to presage wind according to the old rhyme—

"Sea-maw, sea-maw, go sink in the sand;
There's never good weather when ye're on the land."

We refer to it because it is the subject of a discussion in Sir Humphrey Davy's 'Salmonia,' the disputants being Poietes, a poet, and Ornithes, a sportsman:—

"*Poiet.*—I have often seen sea-gulls assemble on the land, and have almost always observed that very stormy and rainy weather was approaching. I conclude that these animals, sensible of a current of air approaching from the ocean, retire to the land to shelter themselves from the storm.

"*Orn.*—No such thing; the storm is their element. I believe that the reason of this migration of sea-gulls to the land is their security of finding food; and they may be observed at this time feeding greedily on earth-worms and larvæ, driven out of the ground by severe floods; and the fish on which they prey in fine weather in the sea, leave the surface and go deeper in storms."

We are not disposed to ascribe the landward movements of gulls to premonitory instinct, and rather conceive that, having had a foretaste of the storm, they have fled from its violence to places where they know that they will find shelter and food. Their movements thus indicate commotion at sea as already begun, and consequently furnish no premonition of the weather to be expected inland. Moreover, it is to be remembered that their visits inland are at a certain season for the purpose of rearing their young. One of their chosen haunts for nidification is among the chain of lakes betwixt Dunkeld and Blairgowrie. This gives rise to a singular traffic. Cart-loads of their eggs, boiled hard, are sold all over Strathmore, and are very palatable.

In a different part of the country—in the vicinity of Kirkcudbright—we were repeatedly witnesses of what seemed to give us quite an original idea as to the origin of the word *gullibility*. Living at a gentleman's seat, our attention was directed to a pair of sea-gulls, which, we were informed, were in the habit of regularly coming once a-year, accompanied by a couple of their young. Perched on the kitchen chimney-top, the cry of "The gulls are come!" sent all the boys of the family to the favourite amusement of feeding the gulls. Each armed with a *bicker of brose*, we pelted the gulls with balls of the solid stuff till we thought they should have burst. Not one of them was ever missed, and the amazing rapidity with which they disappeared down the capacious maw, satisfied us that gullibility, however characteristic of human beings, is specially descriptive of the sea-maw's power of swallowing.

Insects also figure as weather indicators. Incidentally we have already touched upon the sensitiveness of the leech to atmospheric changes. Ants are so sensible of cold that the finest day will not tempt them to place their eggs, or pupæ, at the top of the nest should the air be chill; and it was remarked so long ago as the time of Pliny, that, before bad weather, they are in a bustle to secure their eggs; "forewarned, no doubt, by the perception of an altered temperature," thinks Professor Rennie, who, in his charming 'Insect Miscellanies,' ascribes the sensitiveness of ants, bees, and other insects, to the same sort of feelings which in human beings give warning of bad weather in the form of gouty and rheumatic pains.

As might be expected in creatures so amazingly gifted with instinct, bees are living barometers. When they fly to the hive and none leave it, rain is believed to be near—and with good reason, we think; for Huber records that, while collecting honey in the fields, the working bees are so feverishly afraid of bad weather that a single cloud obscuring the sun sends them homewards.

Spiders are reputed to be so weather-wise that a foreign naturalist* asserts that, when it is wet and windy, they spin only very short lines; but when a spider spins a long thread, there is a certainty of fine

* D'Isjonval.

weather for at least ten or twelve days afterwards." Kirby admits that his observations are in the main accurate, and adds:—I have reason to suppose that a very good idea of the weather may be formed from attending to these insects."

We believe that we were the first, some five-and-twenty years ago, to publish the singular circumstance that, excepting the chief lines, the webs of spiders are sometimes taken down with as much care as they are constructed, and that this is always before rain. Professor Rennie writes:—" We have tried numerous experiments by moving and vibrating the lines of many species, so as to imitate, as nearly as possible, the entrapment of a fly; but in no case have we succeeded in bringing the spider to the spot, because, as we inferred, her eyes always detected our attempted deception." We once were so clever as to cheat a spider. Gently shaking a very small hook, called the midge-fly, in the lowest line of her web, our barometrical friend—whose pre-sensation gave warning of wet—was fairly taken in. Rushing on the hook and grasping it, great was her astonishment. Finding that she should not believe her eyes, she precipitately fled; and no subsequent temptation, though renewed weeks afterwards, enabled us again to boast that we excelled Professor Rennie in angling for spiders.

Though naturalists differ as to the degree in which insects give premonition of atmospheric changes, enough is known to make it very desirable that their relations to meteorology should be systematically studied. Insects are undeniably very susceptible of varying temperature. Kirby and Spence tell us that this susceptibility is probably due to electricity perceived by the antennæ. Rennie is rather disposed to refer it to electricity acting on the hairs with which most insects are beset, and adverts to the fact that bees, which are such electrometers, are among the most hairy of all insects. This surmise is probably well founded, and suggests observation of the fur of animals as likely to furnish important meteorological indications. As this is not noticed by Dr Mitchell, we submit it for his consideration.

The vegetable kingdom also opens up a curious field of investigation, in which the meteorologist will learn much regarding the mutual interdependence of all departments of creation. If light and electricity be so influential in exciting the movements of animals breathing the vital air, plants are equally subject to the same potent agencies, and testify to their influence,—mutely, it is true, but so visibly as to attract the notice alike of the scientific botanist and of the illiterate rustic. In some parts of England the peasants mark the blooming of the large water-lily, and think that the number of its blossoms on a stem indicates the price of wheat per bushel for the ensuing year—each blossom being equivalent to a shilling! We smile at this as superstitious folly; but even philosophy does not disdain "the poor man's weather-glass"—the pimpernel (*Anagallis arvensis*)—and is too wise to despise the weather indications afforded by the shutting of the flowers of the small bindweed, the wood-anemone, the wood-sorrel, and the common daisy, which appears to have derived its expressive name—day's eye—from its sensitiveness to the light.

Such phenomena, as Dr Mitchell notes, are probably determined by the action of light; and the flowers of such plants being shut at ten or eleven A.M., tells of clouds and gloom, and so predicts rain. It has been ingeniously proposed to form a floral timepiece from an arrangement of plants whose periods for opening and closing their flowers are known. The star of Bethlehem expands its flowers about eleven, and closes them at three in the afternoon. The goat's-beard closes its petals at noon, and hence its provincial name of go-to-bed-at-noon. And that light is a chief agent of these changes, is proved by the experiments of Decandolle, made at the Jardin des Plantes, in an underground cellar, illuminated by lamps giving a light equal to fifty-four ordinary wax candles. By lighting these he could cause the flowers of the star of Bethlehem to open at pleasure, and also those of the sea-chamomile, which keeps its flowers closely shut during the night; but he could produce no artificial effect, with the strongest light, upon several species of wood-sorrel, whose flowers and leaves are both folded up at night. With the sensitive plant he succeeded in so completely changing the hour of closure, that on the third day from being placed in the lighted cellar it began to fold its leaves in the morning and open them in the evening.

But not only do very many animals and plants afford popular weather prognostics—these are also found abundantly in the aspects of the sky, in the sun and moon, in the aurora borealis, falling stars, and thunder.

Dr Mitchell, as to these, indicates a very wide field of inquiry, and remarks that "the accusation of fanciful can with most fairness be brought against those prognostics which are associated with the aurora, halos, mock suns, thunder, &c." If he had included among the dubieties those connected with the moon, he would have agreed with most physicists, although it must be avowed some of the most distinguished of these hold that lunar influences do affect the weather. "No observation," says Mr Daniell, "is more general, and on no occasion, perhaps, is the almanac so frequently consulted, as in forming conjectures upon the state of the weather. The common remark, however, goes no farther than that changes from wet to dry, and from dry to wet, generally happen at the changes of the moon. When to this result of universal experience we add the philosophical reasons for the existence of tides in the ærial ocean, we cannot doubt that such a connection exists. The subject, however, is involved in much obscurity." Supposing the connection to be demonstrated, we need not despair of tracing the reason of it.

Thanks to Mr Piazzi Smyth, and other recent observers, we are acquiring new information as to matters lunar, which in due time will, we doubt not, be traced out in their relation to things terrestrial; so that to them we shall not always have to apply the observation of Pliny—"The cause lies concealed in the majesty of Nature."

While hesitating to acknowledge the effect of the aurora on the weather, Dr Mitchell attaches great importance to the following prognostic observed by Professor Christison:—"For a period of between thirty-five to forty years I have never known an exception to the rule, that the first great aurora, after a long tract of fine weather in September or beginning of October, is followed on the second day—and not till the second —about one o'clock on the east coast, and about eleven o'clock in Nithsdale, by a great storm; and that the next day after the aurora is fine weather, fit for all agricultural purposes." Although, then, the same phenomenon may, by different persons in the same locality, be made to predict totally different conditions of the weather—though the aurora has been thought to predict wind by some, and war by others—while the Esquimaux fancy that it is a game which is played by the departed spirits of their relatives—nevertheless, there is room for rational investigation. More than thirty years' observation by one so cautious as Professor Christison is not to be put aside, especially when Fitzroy makes this acknowledgment:—"Among the more experienced seamen who have visited many climates, an opinion prevails that lightning, *the aurora*, meteors, or shooting-stars, are indicative of disturbance in the air, and foretell wind or rain, if not both, in no long interval of days."

Dr Mitchell may also be less sceptical as to halos, when reminded that Sir Humphrey Davy writes thus: —" As an indication of wet weather approaching, nothing is more certain than a halo round the moon, which is produced by the precipitated water; and the larger the circles, the nearer the clouds, and, consequently, the more ready to fall."

"A rainbow in the morning is the shepherd's warning;
A rainbow at night is the shepherd's delight;"

why, of course, he cannot tell. The explanation of Davy is this :—" A rainbow can only occur when the clouds containing or depositing the rain are opposite the sun, and in the evening the rainbow is in the east, and in the morning in the west. As, therefore, our heavy rains in this climate are usually brought by the westerly wind, a rainbow in the west indicates that the bad weather is on the road, by the wind, to us ; whereas the rainbow in the east proves that the rain in these clouds is passing from us."

As to the old faith, that "the evening red is the sign of a bright and cheery day," Dr Mitchell notes that the red after sunset must have a *crimson* tinge, and must last for some time. Davy, however, maintains that the red must have a tint of *purple*, which tint portends fine weather, for this reason :—" The air, when dry, refracts more red or heating rays ; and as dry air is not perfectly transparent, they are again reflected in the horizon. I have generally observed a coppery or yellow sunset to foretell rain."—*Salmonia.*

In regard to underground prognostics—such as the increased flow of water in mines—these are rendered credible by what Arago relates in a singular chapter of his 'Meteorological Essays,' in which he shows that where the atmosphere is tempestuous, there are simultaneously great perturbations in the interior of the earth, and at or below the surface of waters.

For instance, in the Vicentine Hills, when a thunderstorm is preparing, the fountain of Bifoccio, even when it is apparently dried up, suddenly overflows its basin, and fills a wide channel with muddy water. Again, an Artesian well near Perpignan, which at first furnished an abundant gush of water, suddenly stopped. One day when the sky was covered with heavy storm-clouds, there was heard a subterranean bubbling sound, soon followed by an explosion, after which the Artesian well again flowed as at first.

About three miles from the spring of Bifoccio, there is, in the courtyard of Signora Pigati of Vicenza, a deep well, which, at the approach of a thunderstorm, seems in a state of ebullition—sounds issue from it so as to spread alarm among the neighbouring inhabitants. In like manner there is at the Mount d'Or, in Auvergne, a stone basin, called Cæsar's Basin, through which gushes a spring, the increased noise of which is regarded as an infallible indication of the approach of a thunderstorm. Arago, in connection with these and similar phenomena, seems disposed to admit that, when thunderstorms are gathering, water has a tendency to rejoin the clouds, manifesting itself by decided phenomena of intumescence; so that it is certain the meteorologist does well when submitting to careful investigation underground prognostics of the weather.

Altogether, the diversified phenomena to which we have been directing attention, though apparently unconnected, will in all probability be found allied together by some pervading law, the discovery of which will go far in the solution of the mysteries which at present beset the study of meteorology. If there be any who think it derogatory to philosophers to invite them to the humble task of gleaning the few grains of truth scattered through the wide field of popular meteorology, we pray them to remember that the objects on which the illiterate found their prognostics of the weather may be known and read of all men; and that the learning of the wise is never better employed than when investigating the reality of facts which Nature appears to delight in revealing to the senses of the humblest of her children.

HIRUDICULTURE—LEECH-CULTURE.*

COLOSSAL fortunes have been made by rearing leeches. The demand for them is constantly increasing both in this country and in France; and as this can be met only by those possessing suitable rural localities, we are conferring on them a benefit when directing attention to this singular and very remunerating branch of rural economy. Such is the demand for these creatures, that four only of the principal London dealers import about eight millions of leeches annually. The retail price used to be about threepence, but now it is sixpence. In France the demand is also so much in excess of the supply, that the common price is from twenty-five to sixty-five centimes; so that hospitals and charitable institutions are constrained, from economical considerations, to advise medical men to be sparing in the use of such a costly remedy.

This extending use of the leech is intimately connected with a blessed revolution in the practice of medicine. Within the memory of many of us, bleeding was freely resorted to with a most foolhardy forgetfulness of the Mosaic sanitary declaration, "the life of all flesh is the blood thereof." Doctor Sangrado, immor-

* 'La Pisciculture et la Production des Sangsues.' Par Auguste Jourdier, avec une Introduction par M. Coste, de l'Institut. Paris: Hachette et Cie.

talised in 'Gil Blas,' is not a myth, but "a general practitioner," down to very recent times, and whose doings with his lancet were inconceivably mischievous. So widespread was the carelessness about the loss of precious blood, that we are old enough to remember the time when female servants in towns were wont to have themselves bled every spring, although in the highest health—a senseless and debilitating custom, against which we have to remonstrate in the country to this day! But our doctors, like our legislators, being now much less sanguinary in their curative processes, the lancet is superseded by the leech, whose power of bloodletting fortunately does not extend *ad deliqium animi*—the fainting of the patient from loss of blood—the point barbarously aimed at by doctors of the Sangrado school.

Cupping, too, has of late got into disuse; hence the growing demand for leeches. Hence having, in the extension of pisciculture, found grateful food for the healthy, the benevolent M. Jourdier feels called upon, in the name of the general wellbeing, and in the name of the sick poor, to demand that leeches shall be reared and multiplied. The propriety of the demand is evident enough; and that it can be complied with, most beneficially to those who rear them at all events, is demonstrated by a variety of curious details. We shall abstract and translate, so as to give our readers the benefits of his singular information.

Leeches belong to the class *Apodes* of Blainville, and *annelides* of Lamarck, and constitute a small family termed *hirudenées*, or *sanguisugaires* (bloodsuckers). They are all characterised by long flat bodies, and by numerous rings or close articulations, by means of which they move; by a muscular disc, a sort of cupping-glass or sucker situated at the last ring of their bodies; while the anterior part presents another sucker, at the bottom of which is the mouth, which is armed with three triangularly-disposed teeth. By means of this apparatus the animal fixes firmly on the body which it wishes to surround. In the centre of the mouth-sucker are three small but sharp teeth, and near the anterior margin (in the medicinal leech) are ten very small black eyes, arranged in the form of a crescent. At the bottom of the mouth-sucker begins the digestive tube of leeches. Those frequently used in medicine, and so greedy of human blood, easily pierce the skin by means of their teeth. A recent communication to the Société Zoologique d'Acclimatation makes known the existence in America of leeches possessing this valuable peculiarity, that they leave no mark on the skin to which they are applied, so that they must act not by biting but by suction. This curious fact is put beyond a doubt by the experiments made upon himself by Craveri, a learned Italian. M. du Filippi has placed these leeches in a new genus, which he terms *Hæmentaria*, and of which he describes three species, two belonging to Mexico, and one to the river Amazon. The blood which they suck passes abundantly into the œsophagus and the stomach, which is remarkable for its size in comparison with that of the body.

The leeches of commerce are the green (*Hirudo officinalis*) and the grey (*Hirudo medicinalis*). Each of these species presents several varieties, of different degrees of utility, and thus inviting unscrupulous venders to practise their tricks of trade. A good leech is long and flat. The outer skin is of a particular velvety appearance. It moves in water with great activity, and stretches itself out remarkably. Its elasticity is such that it can treble its length, and twine round the finger like a ribbon. The fineness of the anterior compared with the posterior part of its body is also a sign of good quality. Fraudulent dealers mix the different kinds, or gorge them with blood in the proportion of 45 or 50 to the 100, in order to make them weigh more, leeches being sold by weight; so that small leeches, worth 75 francs, are sold for 130. This fraud may be thus detected: The leech, designedly gorged, has a body less prolonged than that of a leech naturally large and empty; it is disposed to present itself under the form of an olive; placed in water, it is dull and sleepy-like; the velvety appearance of the skin is not the same as that of a leech not gorged. When pressed between the fingers, one perceives a reddish reflection, except in the Turkey leech, which does not elongate under the fingers, but the blood accumulates towards the tail, and then is expelled, sometimes in the form of a jet.

The Mode and the Time of Reproduction.—Leeches reproduce themselves by cocoons, experience shows, though some philosophers maintain that the leech is propagated by eggs. M. Borne, in particular, who has for eight years patiently and intelligently observed the phenomena of leech-life, declares that he never saw a single leech produce an egg. The depositing period of the leech cannot be exactly fixed. It commences about the beginning of summer, and lasts a great part of the autumn.

Leeches are hermaphrodites. Each of them possesses a male and a female organ, and during copulation discharges the function both of the male and the female. Copulation begins with the fine weather, and may generally be observed during a part of the month of May. Gestation commonly lasts till July; and the exclusion occurs during August and September. The advanced period of gestation is manifested by an ovoidal yellow swelling in the anterior third of the body, round the sexual parts. This swelling is by zoologists termed the *cincture* (*clitellum*). At the moment of accouchement the cincture is very large and very pale; its epidermis is raised as if about to be detached. The creature twists, opens its mouth, and appears to suffer; speedily there is a compression at each end of the cincture. The leech immediately draws back the anterior part of its body from the ovoid pellicule, which covers the clitellum, as if it were getting out of a hole, and this isolated pellicule becomes the membrane of the cocoon.

The leech withdraws backwards; but before withdrawing deposits in the interior of this bag several little ovules, in the midst of a quantity of albuminous matter. The two openings of the bag soon close, and in their room there remain two round brownish swellings, which afterwards fall, like *opercula*, at the time of exclusion. The cocoon is not yet complete. When complete, the cocoon is about the length and thickness of the first joint of a lady's little finger: it wants the spongy tissue which is deposited upon the membrane, like frothy light slime, of a white colour. The slightest touch removes it. The presence of water is thus always more or less injurious; and the cocoons are almost always deposited either on or in the moist borders of marshes, a little above the surface of the water, and on the plants which deck the moist earth. Having thus, thanks to M. Jourdier, assisted at the accouchement of a leech, we hope our readers are duly grateful; they should be at least; for the phenomena which he so graphically describes are not easily to be witnessed, seeing that by the middle of June almost all leeches old enough to propagate disappear, and hide themselves in the turf of the marshes, artificial or natural, till the time for depositing.

Each cocoon generally contains from fourteen to fifteen threads, or young leeches—at the most, twenty-four to twenty-six; at the least, ten. It is universally observed that the cocoon never contains an odd number.

Such is the nature of these creatures, the rearing of which in great natural marshes, or in marshes artificially made, forms an important branch of rural economy in France. It may be worth the while of some of our Scottish proprietors to attend to it, seeing that it is not merely profitable to those devoted to it, but is also found to have an important bearing upon agriculture. In those communes of the department of La Gironde where hirudiculture has been followed for several years, there has been observed a very visible improvement in the condition of all the inhabitants. Wages, both of men and women, are almost doubled; and they are constantly employed, the one in taking care of the basins and the horses, and the others in catching, cleaning, and maintaining the leeches. The rearing of leeches, according to the mode practised in La Gironde, by necessitating, for the purpose of feeding them, the employment of a great many horses unfit for anything else, gives rise to a kind of fattening so considerable as to deserve to be regarded as one of the most important and useful novelties which have arisen in favour of agriculture. A leech can live in water without food; but it grows lean, and loses its strength and activity. Besides, an ill-fed leech does not reproduce itself, or does so imperfectly. It is of importance to feed these annelides. Their digestion is very slow, and they do not require to be fed more than twice or thrice a-year. M. Borne does not go beyond that, and he takes three years, at the most, to *make* a leech. The people round Bordeaux profess to make a leech in eighteen months, and even in a year —an assertion not easily admitted by one who has paid some attention to leeches; and yet the fact is demonstrated in the marshes of M. Franceschi, one of the breeders in La Gironde. Different modes are resorted to in order to produce rapid growth; but the best mode of feeding, and that which alone is to be employed if we wish to carry on operations on a great scale, is to gorge the leeches by living animals. Warm blood drawn from the veins of an animal, through the skin, is admitted to be that which best suits the leech, and is followed by the most remunerative result. Throughout La Gironde, horses, asses, mules, and cows are made to enter the marshes as food for leeches. The disturbance caused by their entrance attracts the leeches, which come forth from their retreats, and fasten and gorge themselves on their living prey. Bridges lead to the different divisions of the ponds into which the animals are driven, and are immediately covered with vast numbers of leeches, which never leave their prey until completely gorged. The victims are then taken back to a poor pasture, where they painfully endeavour to regain the blood which they have lost. They are thus tortured five or six times a-month. It is shocking to see them at these periods, which are usually from the beginning of April to the 15th June, and then from the beginning of October to the 15th November.

Horses are much preferred to all the other animals we have mentioned. Cows feed leeches well, provided they can be got into the water above the knee; otherwise, with their rough tongues they easily shake off the leeches. Asses would be esteemed on account of their docility, but their narrow feet sink too deeply in the bottom of the marshes. But the unhappy donkeys are not permitted thus to escape. M. Franceschi, of Bordeaux, is the inventor of what may be styled drawers for donkeys; their legs being enveloped in these, the drawers are filled with leeches, in proportion to their size, and in the hope that the health of the animals may not suffer. We are glad that this hope is not fulfilled, and that in 1854 the ingenious barbarity of M. Franceschi resulted in the death of all the donkeys so maltreated. M. Jourdier's sensibility is pained by this brutal mode of feeding; and yet, while sympathising with the suffering animals, he gives the advice that they shall always be in good health—*for the sake of the leeches!* Just as the friend of a cannibal might, from anxiety about his health, counsel him only to feed on healthy men!

There is need, however, for M. Jourdier's advice, for in the horse-market for leeches at Bastide all kinds of rascality is rife, in order to get rid of disabled or diseased animals. The horse-dealers of Bordeaux are at least a match for those of Paris; and their efforts to hide defects are always culpable, sometimes comical, but generally distressing. M. Jourdier's excellent friend, L'Héritier of the *Pays*, was so saddened by what he saw that he had no wish to visit the marshes.

A less disgusting mode of feeding leeches is that followed by M. Borne and Dr Sauvé, which has the additional recommendation of being economical. In the first days of spring, when the leeches are moving briskly in the marshes, and when they rush in multitudes to the place where the water is disturbed, we may conclude that they are in want of food; for leeches rush out when a noise is made, only in hope of living prey on which they may fasten. Their capture is effected thus: In the large marshes of La Gironde, men and women, protected by large boots well greased, go into the marshes to fish for leeches. In the left hand they hold a bag of very thick linen, and, stirring the water with their feet, they seize with the right hand the leeches which rush out in hope of prey. The usual way of transporting them is in sacks of thick linen, placed side by side in a long square basket, surrounded with straw in winter, and with moist reeds in summer. This mode proves fatal to vast numbers. M. Borne uses a box, varying in size according to the number of leeches to be transported. For six thousand he uses a box of a little more than three feet in length, and divided into six compartments. At the bottom of each of these he places turf or reeds, and over these moss. On this moss he places a bag containing a thousand leeches, which are likewise covered with moss. And so with each bag in succession, in order to avoid the slightest shock. They can thus be carried long distances with a trifling mortality. A still more useful invention is the domestic marsh of M. Meeus. This is a box filled with river water, turfy earth, and aquatic plants, in which the leeches live as in a natural marsh, deposit their cocoons, and change their outer skin. Two of these boxes—one for leeches not used, the other for leeches which have sucked—are found to be most useful in public hospitals. Those desiring to preserve leeches will do well to remember that the water employed must be river or rain water, but never water from a well, nor spring water, as it issues from the earth; that vessels of earthenware are preferable to vessels of glass or glazed earth; and that darkness is fatal to leeches.

But to return to the mode of feeding followed by M. Borne; here it is:—Placing the leeches in little bags, he plunges them in a bath of warm blood flowing from the veins of an animal. The bags are of flannel or of fine linen, for large and moderately large leeches. The very small leeches are put in bags of muslin or fine flannel, and carried to the shambles. When the ox, calf, or sheep is bled, the blood is beaten in order to destroy the fibrine, and hinder the formation of clot, and then the bags are plunged in it. The tissue in which the little creatures are shut up serves as a point of attachment for sucking and affords the means of occasionally looking at them, to see if they are gorged. The larger leeches are left in this bath of warm blood for five or six minutes, those of middle size for about ten minutes, the small ones for a quarter of an hour, and the very small ones for half an hour. They are then taken out, washed in warm water, and then placed in cold water to be taken back to the marsh.

M. Sauvé also uses bags to gorge his leeches, but in some respects his plan differs from that of M. Borne. He carries the blood in cans with double bottoms, which are filled with boiling water to prevent cooling; and then, like M. Borne, he plunges into the vessels full of blood the leeches in bags of linen, or preferably of woollen. This warm blood, which retains its fibrine, is, M. Sauvé maintains, much superior to that which has lost its heat and its fibrine.

It is a kind of satisfaction to know that these blood-thirsty creatures are preyed upon by many enemies, and make dainty morsels to sand-mice, water-rats, moles, the larvæ of the dragon-fly, eels, herons, and, above all, wild-ducks. Well may leech-breeders quake when told that in twenty-four hours 200,000 leeches have disappeared in the greedy maws of a flock of wild-ducks.

The details we have been giving are certainly curious, but their disgusting nature will not repel those who reflect what profit may be made from those marshes which are so numerous in this country, and hitherto of almost no value. M. Béchade, a small farmer in the neighbourhood of Bordeaux, has become a millionaire by transforming poor marshes, for which he could hardly pay a rent of 300 francs, into magnificent enclosures for leeches, now let for 25,000 francs. M. Jourdier refers to a Parisian capitalist who has embarked in this species of industry with the satisfactory result of a revenue of fifteen for one; that is to say, a leech which then (April 1855) cost twenty-five centimes, produced on an average fifteen leeches a-year, which could be sold at the same price—or, say, three francs. Deducting at the most five centimes of expenses, there remains a gain of two francs twenty-five centimes, which is enormous when the operation is on a large scale. It is, therefore, credible that a marsh of forty-eight hectares should let for 250,000 francs, and that enormous fortunes should have been made by this new species of rural economy, which is alike useful to the public and beneficial to the private interest of those by whom it is prosecuted. We shall, therefore, not be surprised at its introduction into this country. We earnestly hope that recourse will be had to a less revolting mode of feeding leeches than we have had the pain of describing: this may possibly be effected by furnishing them with abundant supplies of their ordinary food, which appears to be tadpoles, aquatic worms, and larvæ.

A correspondent asks, "Would leeches thrive in this climate?" In this country only two kinds are reckoned fit for medical purposes—viz., the "brown leech" and the "green leech"—the former found sparingly in Great Britain, but abundantly in northern and central Europe. That it will thrive in this country we doubt not. Leeches abound in India, and we should not despair of receiving them from so distant a region. France imports leeches from Hungary, Turkey, and Syria, besides exporting them to Brazil, Chili, and Peru.

We conclude by informing our readers that the susceptibility of the leech to changes in the atmosphere makes it a true weather-prophet; and Mr Merryweather, taking advantage of this peculiarity, has constructed an instrument in which the movements of a leech set a bell a-ringing, and so announce a coming storm.

MARITIME PISCICULTURE—LAGOON OF COMACCHIO.*

There is urgent need for the application of science to the important branch of natural history which relates to fishes; the folly and the greed of man having counteracted the beneficence of the Creator, and occasioned an alarming diminution in the numbers of even those species of fish the prolific energies of which might be thought inexhaustible. A sturgeon has been computed to produce 7,653,200 eggs; and yet this is a fish by no means superabundant. If all the ova of a herring were fertile, it has been calculated that it would not require more than eight years to fill up the bed of the ocean, and that if our globe were covered with water, it would soon be unable to contain this single fish! And yet, but for the restraining interference of legislation, even the herring would ere long bear witness to the power of man in sensibly reducing the numbers of a fish which may be regarded as the type of abundance.

In this country the depopulation of our rivers has long been lamented. It is the same in France; fresh-

* Voyage d'Exploration sur le Littoral de la France et de l'Italie. Rapport à M. le Ministre de l'Agriculture, du Commerce, et des Travaux Publics, sur les Industries de Comacchio, du Lac Fusaro, de Marennes et de l'Anse de l'Aiguillon. Par M. Coste, Membre de l'Institut, Professeur au Collège de France. Paris: Imprimerie Impériale.

water fish are rare, and consequently so dear as to contribute little to the food of the labouring classes. It is nearly the same with sea-fish. M. Jourdier declares that the maritime population deplores their disappearance, and that the produce of the sea-fishings barely suffices to supply the demands of Paris. "Where," he asks, "are those magnificent fish formerly so common? They have disappeared, or, when they occasionally appear, are sold at exorbitant prices." In truth, the position of Paris, in regard to the supply of food, is every now and then alarming. Bread and butcher-meat are distressingly dear, so that we are not surprised at the attention of the Government being seriously turned to *pisciculture*, and that great encouragement should be given to the interesting and important researches of M. Coste. At his suggestion the piscicultural establishment at Huningue—that vast laboratory intended for the study and perfecting of the modes of artificial propagation—has been incorporated into the Commission for the Construction of Bridges and Highways. In passing into the hands of this comprehensive and able body, such an impulse has been imparted to it that it has already become a powerful instrument for the artificial propagation of the best kinds of fish; and it is making provision to commence on a grand scale the restocking of the rivers of Europe.

The fish to the introduction of which the greatest importance is attached is the salmon of the Danube. It deposits its eggs in May and June, in the tributaries of that river; its flesh is white and delicate, and it is never sold in the market at Munich at a less price than 1s. 3d. per lb., although the supply is very abundant. This noble fish attains three times the weight of the salmon in the same length of time, and sometimes weighs 240 lb. "The young Danube salmon we reared two years ago at the establishment of Huningue, have now attained," states M. Coste, "the weight of five lb., and those which were hatched this year had in August attained the same size as salmon and trout hatched five months previously. The acclimatisation of this fish is the more easy, as it never quits fresh water. In some parts of Germany, especially in Hungary, it thrives along with the carp and the pike in the same lakes. The temperature favourable to these fish agrees also with the Danube salmon; so much so, that it attains the weight of from thirteen to fifteen lb. in three or four years. It can even be reared in private fish-ponds. I saw at Munich, in the pool belonging to M. Schissel, fisherman to the King, some very large Danube salmon, which were fed there for the purposes of daily consumption."

We refer to these details because desirous to attract attention to the ease with which our lakes and rivers may be stocked with valuable kinds of fish with which we are not acquainted. When it is demonstrated that their ova can be safely transported to great distances, that the expense of artificial rearing, when this is resorted to, is trifling, and that the pecuniary return is rapid and certain, it is astonishing that the proprietors of our lakes and rivers have not recourse to such a method of at once benefiting themselves and the nation by such an addition to the food of the people.

M. Heurtier, Director - General of Agriculture and Commerce, observes in reference to *pisciculture*—" In confining itself to the artificial fecundation of fresh-water fishes, the problem appears to me to be only partially solved. It is indeed equally important to extend the application of this discovery to sea-fish. Now, especially when our great lines of railway have to a certain extent destroyed distances, sea-fish will be readily transported to almost all towns, even the most distant; to only a few will they be carried in a preserved state. While endeavouring to multiply sea-fish, crustacea, and molluscs, it will then be equally useful to inquire into the best modes of preparing and preserving them." M. Heurtier then proposes that M.

Coste shall visit the numerous lagoons in the south of France—the waters of which, generally salt, but occasionally mixed with fresh water, will, he observes, "furnish the means of interesting fecundations and naturalisations, and, if the prognostications of science be verified, might be converted into valuable reservoirs of all kinds of fish." He also proposes that the distinguished naturalist shall visit the lagoons of the Adriatic, near the mouths of the Po, the Adige, and the Brenta, and, above all, Comacchio, where, from time immemorial, and on a great scale, fish of excellent flavour are preserved. M. Persigny, then Minister of the Interior, having sanctioned these enlightened views, so interesting to science, and so important to national economy, M. Coste repaired to the assigned scene of his labours, and amassed those details which are now published with the view of aiding the formation of similar kinds of industry along the shores of France.

Science being the common heritage of nations, whence each people may borrow those usages which suit it best, the publication of M. Coste's important inquiries is a common boon, for which the thanks of the world are due to the enlightened liberality of the French Government. To a maritime people like us, whose colonies, moreover, comprise the most extensive fisheries in existence, such a work is especially interesting: we therefore desire to make known its contents, believing that they will suggest new modes of multiplying, capturing, and preparing several kinds of fish, the commercial value of which is very great.

This paper will be restricted to a description of the singular fishing community of Comacchio.

When the traveller has passed through fertile and populous Lombardy and arrived at Ferrara, a few hours' walking in the direction of the Adriatic brings him suddenly into the midst of a flat, sandy, desolate region, where reign silence and misery. "Its few inhabitants," M. Coste informs us, "have so little intercourse with neighbouring countries that not a single public conveyance conducts to its territory, although the road by which it is traversed be the only one to the most curious, but perhaps least known, industrial colony anywhere existing—that of Comacchio, the population of which, at the time when the barbarians chased before them civilised nations, came, like the founders of Venice, to take refuge in the immense marsh where, for centuries since, they have employed themselves in transforming it into the veritable means of cultivating the ocean, and where their ingenious industry attracts the young fish hatched in the Adriatic, and where they gather them, when grown, by proceedings as rational as those of the agriculturist who sows the ground and gathers the produce." Indeed, they themselves are so struck by the similarity of their processes to those of agriculture, that they have always termed the basins among which they pursue their avocation *fields*, as if they were true cultivators of the soil; the annual ascent of the fish being the sowing of these fields. "Less favoured than their neighbours at Venice, and unable from their inferior position to aspire, like her, to commercial supremacy or the aggrandisement of conquests, they directed their genius to the construction of an admirable system of embankments, made of the mud of lakes, strengthened by fragments of shell-fish, intersected by numerous sluices connecting together well-managed canals, which, by affording access to the waters of the Adriatic and of the rivers bordering two sides of the lagoon, enable them to have the entire or partial command of the lagoon as thoroughly as if they had to do with the simple apparatus of a laboratory. Here these simple people have remained submitting to the monotony of a rigid discipline, and losing their sleep when the storm lashes the lagoon and raises the waves, satisfied with moderate wages and a daily allowance of fish."

Like other fishermen, they are clannish, remarkably chary of new alliances, and no less averse to strangers settling among them. The women are remarkably handsome, the men very athletic; and the veils of the one and the long tasseled bonnets of the other recall their Eastern extraction, and favour the idea of their being a remnant of one of those Grecian colonies which, during the Roman rule, peopled the shores of the Adriatic. They are now devout subjects of his Holiness the Pope, and their abundant supply of fish to the States of the Church, as well as to Venice and other great towns of Italy, enables the people easily to endure the fasts prescribed by ecclesiastical authority. The lagoon which yields so rich a harvest is situated between the mouth of the Po and the territory of Ravenna, and forms an immense marsh, 140 miles in circumference and from 3 to 6 feet deep, and separated from the sea by a narrow belt of land. It is bounded by the two rivers Reno and Volano, which enclose it in the form of a delta. Among the numerous islands with which its surface is dotted there is one, long and narrow, in the very centre of the lagoon; and here these industrious fishermen laid the foundation of a town now containing 6661 inhabitants, and consisting of a single street of houses of only one storey, on account of the violence of the winds. The idea of transforming this lagoon into an immense field, from which might be gathered an annual harvest of fish, was suggested to these industrious fishermen by the discovery of that special instinct which impels certain kinds of fish, shortly after being hatched, to ascend water-courses in countless numbers, and to regain the sea after they are fully grown; a curious phenomenon annually recurring in every quarter of the globe at the mouths of channels discharging themselves into the ocean, or which from it receive their waters. At the mouths of these channels myriads of very small transparent fish rise to the surface about the usual periods, and advance in masses more or less compact, and sufficient to replenish all the waters of the earth, if they were protected from destruction, or conducted, as at Comacchio, to reservoirs where they might be converted into harvests of alimentary substances. This animated matter, when carefully examined, is found to be composed of thread-like animalcules—the young of eels, soles, plaice, mullets, &c. These periodical migrations are termed *montée*, and last from February to April or May, according to the temperature, or difference of climates. Three-fourths of our terraqueous globe consisting of water, the quantity of alimentary substance derivable from fish is inconceivably great, if communities would only wisely avail themselves of the bounties of Providence. But they either neglect or recklessly waste them, so that it is by no means rare to hear of the young of the salmon and of the herring being caught in such quantities as to be converted into manure. The proceedings at Comacchio are a standing protest against such wasteful follies. The whole lagoon has, with incredible labour, been intersected by dykes and sluices and canals communicating both with the sea and with the rivers Reno and Volano, so as to form a hydraulic apparatus, the whole or any part of which can be put into operation at pleasure. M. Coste declares that its productiveness is unlimited, if to their ancient usages these fishermen were to add those of artificial impregnation, with which he has made them acquainted.

Five hundred men are employed in this species of industry, and are subjected to the most rigid discipline; living together in barracks, and at common tables, the staple food being their daily allowance of $1\frac{1}{2}$ lb. of fish, chiefly eels, grilled and seasoned only with their own fat, and which M. Coste found not only very delicate but also very digestible.

The *seeding* of the lagoon commences every year, on the 2d February, by the opening of all the sluices, this being the period when the newly-hatched fish begin their annual ascent. The Papal Government at this season rigorously prohibits the use of small-meshed nets on any part of the shore connected with the lagoon, and will not even permit those with a wider mesh, except at a certain distance from shore, in order that no damage be done to the ascending fry. When the head of these long *seed-columns* begins the ascent, all the rest continue to follow, unless some unforeseen incident break the endless chain; a calamity which is carefully guarded against by the removal of every obstacle. About the end of April all the sluices are again closed, and the lagoon is reconverted into a basin hermetically sealed, within which the aquatic flocks are henceforth retained within their respective compartments until reckoned fit for sale. The soles, generally lying upon the mud, chase the worms and insects on which they feed; the mullets, intrepid voyagers, move about perpetually in search of animals weaker than themselves, but living mainly on the marine plants, or the organic matters with which they are covered. These singular creatures, when aware of being caught in the divisions of the labyrinths, mount to the surface, put their large heads out of the water, and with a noisy rush follow the fishermen on the banks, as if expecting them to be their deliverers; a sight which recalled to M. Coste the words of Martial " *Nomenclator mugilem citat notum, et adesse jussi prodeunt mulli.*"

The *aquadelle*, a fish of very small size, forms within these basins innumerable banks, and appears to be the prey appointed for their carnivorous population. The eels in particular are their cruel assailants. In order to chase them, they assemble in those places where there are falls in the water, the favourite locality of these tiny fishes, and, darting upon them, intertwine them in voluminous knots. So great is the fury which animates them that they can be diverted from their object neither by boats passing above them nor by the approach of nets. They persevere till their ravenous appetite is satiated, when they bury themselves in the mud, where they lie till hunger drives them forth from their retreats.

The basins of the lagoon must be a delicious place to the eels and the other species by which it is inhabited, seeing that they enter it at their birth, and actually do not seek to leave until grown up, when, their reproductive instinct impelling them to return to the sea, advantage is taken of this to reap the harvest at a time when they are so large as to be eatable. The fishermen of Comacchio cannot speak precisely as to the length of time required in order that the eel may attain its greatest size, some saying five years, and others from eight to ten. But since M. Coste has called attention to the advantages of introducing the breeding of eels in the ponds, lakes, and marshes of La Pologne, it has been repeatedly proved that young eels, placed in basins and properly fed, will, in from four to five years, attain the weight of four or five pounds; so that one pound (or 1800) of young eels will in this time produce 6562 lb. of flesh, worth at least £41. M. Jourdier states that young eels, warranted to arrive safe at their destination, may be purchased by those engaged in pisciculture for forty francs the thousand; so that one may judge of the enormous profit of rearing eels. We are quite aware that in this country the appetite for eels requires to be created before we can hope to see an eel-market. Though eaten in England, they are held an abomination in Scotland; and we verily believe that most of us would as soon dine off the sea-serpent as partake of eels grilled *à la Comacchio*.

If the fishermen at Comacchio be unable to give precise details as to the growth of eels, it is not so with mullets, which, next to eels, constitute their great harvest. Owing to special reasons, a portion of the young of this fish is kept in a pond for a whole year, during which it is possible to note their increase from time to time. At the age of one year four mullets weigh a pound, whereas, when introduced into the lagoon, it required about 6000 of them to make a pound; so that 1 lb. of mullet-fry is in twelve months transformed into alimentary substance weighing above 1500 lb. M. Coste truly observes that "agriculture produces no such enormous and inexpensive harvests: its products are obtained at great expense; those of pisciculture, on the contrary, are developed without the necessity of having recourse to those costly methods which absorb the greater part of the revenue."

This naturally leads us to the description of the fishing harvest at Comacchio—the great event of the year, and the commencement of which is celebrated by a religious solemnity, as other States open those assemblies which preside over their political destinies. The fishermen, prostrate in their chapels, implore the divine blessing on their labours by addressing their prayers to St Gratian, the patron of the colony; and when their harvest-fields have been blessed by the officiating priest, they proceed to open the sluices which admit the waters of the Adriatic into the lagoon, all the outlets from which are now provided with labyrinths. This accession of fresh water in every part excites everywhere the migratory instinct; and the fish reascend those currents which lead to the Adriatic, but are obliged to follow a course leading to the labyrinths. They pass through all the windings till they arrive at the last compartments, where they sometimes accumulate in such numbers as to fill them to the brim.

Dark rainy nights, when the cold north wind raises the waves of the sea and the lagoon, are the most favourable for these operations, and are looked for by the fishermen as anxiously as the agriculturist looks for the sun which is to ripen the fruits of the earth. "They doubtless regard the tumults of nature as a manifestation of the highest harmony, which they term *order:* and when the storm is unroofing their houses, they cry out with delight, 'Order! order!'* just as other people would say, 'A fine day!'" When night has come, all are at their posts round the labyrinths, watching in profound silence, in order not to alarm the fish which are entangling themselves in the insidious routes prepared for them. They wait till the chambers are full, and the moment they perceive them overcrowded, hasten to relieve them; for, if overcrowded, the eels might become restive, and break through the partitions which confine them. The eels are therefore removed in circular baskets, suspended in the water between a couple of poles, and are kept alive till needed. As Bonaveri relates that, during a furious storm on the night of the 4th October 1697, there were caught in the lagoon 200 baskets of eels and more than 1000 baskets of fish, the whole weight must have exceeded 645,040 lb. In the basin of Caldirolo, Spallanzani saw taken, during a single October night in 1792, 12,800 lb. of fish, "which," adds that great naturalist, "is little in comparison of one capture of 40,000 lb. weight, and of another of 19,200 lb., in the same basin and in the same space of time.

When any single valley in one night makes a capture weighing 48,000 lb., a cannon announces the event to the town, in order that, in the middle of their slumbers, its inhabitants may receive the good news. On such happy occasions it is the custom for strangers, ladies of distinction, the family of the farmer-general, and the bishop himself, to visit the valley privileged to be the scene of such extraordinary captures. The commander does the honours of his domain, exhibiting to all the rich harvest, the produce of which the fishermen convert into an ample feast. Eel-broth, so esteemed by the Greeks, but somewhat vulgarised in the eyes of M. Coste by an admixture of *cabbage*, is the leading dish at the hospitable table. Then follow boiled mullet, plaice, dories, and then again the finest eels, and all the

* We suspect that M. Coste has mistaken *ordine*, which certainly means "order," for *all'ordine*—"make ready."

other sorts of fish, grilled or roasted on the spit; the enjoyment being heightened by the wine of Bosco-Eliseo.

M. Coste notices a peculiarity which ought to be known by those engaged in fisheries. The migratory progress of the eels is instantly stopped by the rising of the moon. Their repugnance to move during moonlight, which does not arrest other kinds of fish, has taught the fishermen to use a device in order not to be incommoded by their excessive numbers. When they wish no more of them for the time, they light fires on both sides, and the creatures stop: these are extinguished as soon as the chambers are emptied, and the migration is resumed till daylight.

This harvest lasts for three or four months, and some idea of its amount may be formed when we state that, from 1798 to 1813, the average annual weight of the fish was more than 1,935,120 lb.; from 1813 to 1825 the average was about 1,612,600 lb. But in 1825 the mortality among the fish was so great, in consequence of an accident, that the produce of the fishery fell to about 645,080 lb., and for the eight subsequent years this was the average. Setting out from 1833, and in spite of three successive accidents which occasioned the death of fish weighing more than 9,676,600 lb., the produce remounts towards its former level, but at present the average does not exceed about 967,560 lb. And yet these enormous figures are believed to represent only one-half of the actual captures, so great is the loss sustained owing to the impossibility of efficiently watching so large a body of water. The actual produce may therefore be estimated at 3,970,240 lb. And that this estimate is not exaggerated is proved by the astounding fact, that in a single day the mortality among the fish is sometimes so great as to bring to the surface a greater weight than this of dead fish. M. Coste adduces, as his authority for this statement, M. Ducati, who saw them buried.

The causes of these disasters are sometimes excessive

heat and sometimes excessive cold, to the effects of which the lagoon is exposed owing to its being only a few feet deep. Such a calamity occurred in 1825, when, in order to escape such a plague as decimated them in 1671, the colony was obliged to dig immense trenches, in which they buried these masses of flesh, and burned them with quicklime.

The disposal of such immense masses of a perishable species of food is effected by selling the fish fresh, or prepared for the market. The preparation is of three kinds, carried on simultaneously in the same workshops, but each forming a special department of labour, and requiring a special class, as it were, of manipulations. The first method is termed by M. Coste that of "cooking and acetic salting."

The kitchen, the active centre of the manufacture, is an immense place, with several chimneys like those met with in buildings of the middle ages, and in which were burnt trunks of trees. In front of the chimneys are six or seven spits, like the cross-bars of a window, and below these is a little canal for receiving the oil, exuding from the roasted eels, which is preserved for other processes. The poor eels are treated with the most dexterous manipulation. A workman, seated before a block, with a hatchet in his hand, seizes them one by one in a basket on his left, and having with astonishing quickness chopped off the head and tail, which are the perquisites of the poor, he makes of the body one or two pieces of equal size, which he throws into an empty basket on his right.

Each of these pieces receives a slight trimming, in order to facilitate the operations of other workmen, who, with equal celerity, transfer them to the spits. Only the larger eels are thus decapitated and truncated, their size rendering it difficult to spit them living; but the smaller are devoted to this torture after one or two trimmings, to facilitate their being entwined round the spits. This mode of cooking eels on the spit, either whole or in segments, can be traced back to the old Romans; for in two paintings discovered at Pompeii, on the outer pillar of an inn, the figures on the signboard are in the one a whole eel twisted and spitted, and in the other three pieces on one spit.

Another curious operation at Comacchio is the frying, with eel-fat and olive-oil, in frying-pans at least $2\frac{1}{2}$ feet in diameter, of mullets, dories, soles, small eels, aquadelles, and in general all kinds of fish which cannot be spitted. M. Coste is considerate enough to state that *they* are not thrown in alive. Our philosopher, moreover, has an eye for something more than spits and frying-pans for the finny tribes. We find him noting the youth and beauty of the Comacchian damsels, whose delicate fingers so nimbly wrap in wheaten flour the tiny aquadelles, which they form into *bouquets*, and whose merry tongues relieve their toil with cheerful tales.

The barrels in which the fish are packed are of all sizes, from the huge tun down to those termed *zangoli*. When packed, they are sprinkled with a particular mixture of salt and vinegar, the vinegar being of the strongest, and the salt not white, but grey and earthy, on account of its supposed property of tempering the too great acidity of the vinegar, or possibly because it is cheaper. It is also supposed that the earthy matter aids in warding off putrefaction. To about 124 lb. of vinegar they add about 13 lb. grey salt, if the mixture be for large fish. Great care is taken in order that the fish may be thoroughly saturated, and that the barrels be impervious to air.

The second mode of preserving is that of simple salting, which has given rise to the process termed *basto*, which is applied to eels and mullets; the other sorts not being sufficiently abundant to be formed into stacks. These are always squares, varying in dimensions according to the quantity of fish they contain, and consist of alternate layers of salt and eels, crossing each other like wood piled in a timber-yard. When this stack of flesh has attained the required dimensions, it is covered with a last layer of salt, and surmounted by a weighted plank, which presses the rows closer, and hinders their being penetrated by the air. In about a fortnight, when the eels are saturated with salt, the *basto* is said to be ripe, and the stack is taken down, in order that the eels may be put into the different-sized barrels.

This process is mostly resorted to during seasons of great mortality. In fact, when, either from excess of heat or of cold, the fish are seen rising to the surface of the lagoon, they are immediately gathered in heaps, and subjected to this operation, as the most expeditious and the least expensive.

The third mode of preparation is that of "desiccation," which always begins with an operation termed *salamoja*. The salting, which is to be followed by drying, is effected by immersion in the *salamova*—that is, the liquor exuded from the *basto*, and the baskets in which mullets are salted. The fish are plunged into a basin of this concentrated liquor, where the larger of them remain from eight to twelve days, and the smaller from four to six. An immersion of five or six hours is sufficient for the *aquadelles*, which are then dried in the sun. Thus dried, they become food for the people, who prepare them in a very simple way: they first roast them on the hearth-stone, and then finish the cooking under the hot cinders. The mullets, soles, and dories become so hard when dried, that they cannot be eaten unless steeped for a whole night in soft lukewarm water.

The best as well as the secondary eels may also be treated in this way, but they must be thrown into the basin *alive!* for here, as at the spit, this species is subjected to prolonged agony. "It is pitiful," observes M. Coste, "to see them swallowing and rejecting the boiling liquor, exhausting themselves in fruitless efforts to escape, and writhing on the surface, as if suspended over the gulf; the length of their suffering in which is the very condition necessary for the success of the operation. This cruel necessity cannot be dispensed with without the risk of this species of industry losing the fruit of its toil. If immersed after death, the entrails of these creatures, absorbing too little salt, would corrupt, while the outer flesh would present a deceitful appearance. If established usage was departed from, there would be a risk of seeing at table superb eels, which, when opened, would exhale a fetid odour; and therefore the merchants consider immersion after death a fraud, against which they protect themselves by opening and smelling the mouths of the large eels which they purchase."

Now, as our philosophical friends in France would say, "here we have to do with a question of high morality." Yes, truly. God has given unto omnivorous man "all things richly to enjoy." But surely there are restrictions on the mode of that enjoyment. The merciful Being, who of old time so sternly forbade a Jew to be cruel to the lower animals, cannot be rationally supposed to look with complacency on a Christian torturing fish because the torture fills his purse, or helps to stimulate his luxurious appetite.

When taken out of the boiling liquor they are *embouched*, as M. Coste phrases it; or, to speak intelligibly, their mouths are opened, and, by means of a wooden ram-rod, powdered salt is introduced into their intestines. They are then washed in lukewarm water, fastened together in pairs, and suspended from long poles under the floor of the kitchen, or of some heated chamber. When thus dried they assume a bronzed appearance, which causes them to be termed *smoked*, a term likewise applied to all fish prepared by being dried, although smoke is not employed at all.

We have thus given our readers some idea of the valuable contents of M. Coste's splendid work. We have still, however, a great deal to say in elucidation of its importance as an addition to those works which treat of alimentary substances.

MARITIME PISCICULTURE—OYSTER-CULTURE.*

CONTENTMENT being an eminent Christian grace, we are bound, no doubt, greatly to admire poor Goldsmith, in his Grub Street garret, trying to persuade the world that

> " Man wants but little here below,
> Nor wants that little long."

This may be true of many an individual frequenter of Parnassus hill, whose summit, being in cloudland, is far above the line of profitable cultivation. Of man in general—of man working hard and having a prodigious appetite—of man working little, and needing to have his stomach cheated into the exercise of its functions by the seductions of an ingenious gastronomy—of man " in populous city pent," whose aggregate omnivorousness is astounding—the starved poet's saying is glaringly untrue. " Man needs but little!"—no man in his senses will say *that*, after seeing steamships and vessels from every quarter of the globe unloading in

* Voyage d'Exploration sur le Littoral de la France et de l'Italie. Rapport à M. le Ministre de l'Agriculture, du Commerce, et des Travaux Publics, sur les Industries de Comacchio, du Lac Fusaro, de Marennes, et de l'Anse de l'Aiguillon. Par M. Coste, Membre de l'Institut, Professeur au Collége de France. Paris: Imprimerie Impériale.

the London docks their multifarious cargoes; nor will he say it after seeing, in Christmas week, railway trains rushing into the metropolis laden with the cattle from a thousand hills, and with fowl of every wing;—all soon to disappear in the capacious maw of the million-peopled city!

The truth is that man is needing a great deal more every year that he goes on peopling the earth, and advancing in the refinements of civilisation; and how to provide increase of food for the increase of eaters is in our day the great problem to the solution of which social science is being applied with the greatest zeal, and, let us be thankful, with the most encouraging success. We are in no danger of starving in God's beautiful world, if we will only use our faculties of mind and body in providing food convenient for us. Our difficulty lies in determining what *is* convenient: we are saucy, and turn up our noses at what is really excellent, palatable, and rich in all elements of nutrition. Have we not demonstrated the satisfaction of all truth-loving souls, as we had fondly hoped, that hippophagy, or eating of horse-flesh, is highly to be commended? And yet, in all broad Britain, what horse has been eaten? Not one, we fear. What we know is this: certain ladies look upon us with an aversion which might be justifiable if we had been advocating the child-eating proposal of the facetious Dean Swift.

And we also fear that our last article on maritime pisciculture, in which we urged the profit of rearing and the pleasure of eating *eels*, has been equally disgusting to all save an enlightened few who, in foreign parts, have got rid of their Scotch prejudices. Now, however, we hope that we are about to treat of what will excite an appetite for hearing, to be followed in due time by the literal inward digestion of the savoury subjects of this communication—oysters, namely. And yet, as to the former, an unhappy squeamishness of the inner man hinders some enjoying the " natives," either in the simplicity of bearded rawness, or in the savoury comeliness of patés. Our youthful reminiscences lead us to be compassionate to the idiosyncrasies of those stomachs which rise against raw oysters. The first dish of them presented to our inexperience filled us with horror, as we more than suspected that they were dead men's eyes! A companion, more adventurous, having introduced one into his mouth, was seized with such difficulty of either eating or swallowing it, that he actually carried it home to his mother between his teeth, and from her received an explanation of the nature of the mysterious morsel.

The subject of ostreo-culture is now embraced within the interesting field of inquiry to which we apply the term pisciculture; and the results are so remarkable as to be worthy of the attention of all interested in maritime industry. In regard to oysters, also, we have to deplore that diminution of their numbers, of which we have such growing experience in the case of salmon, and other valuable species of fish. The reason of the enormous loss thus sustained is not to be sought for in the lessened energy of the reproductive powers of these tenants of the waters, but in the senseless greed of man, destroying them by millions while their generative functions are in exercise, or before they have attained maturity. Oyster-beds, once famous for the abundance and quality of their products, have been depopulated alike on the shores of Great Britain and of France. Of the consequent injury to the national wealth of both countries some conception may be formed when it is known that, from the island of Jersey alone, 200,000 bushels of oysters are annually exported; and 250 boats, 1500 men, and 1000 women and children, are employed during the season. The artificial rearing of oysters at Marennes furnishes annually fifty millions of oysters, averaging, according to M. Coste, three francs per hundred; and thus yielding the enormous sum of two millions of francs. The oyster is therefore of importance as food, and as an element of commerce,—entire maritime communities being dependent upon it for their prosperity. And when we reflect on the impolicy of doing anything which shall have the effect of turning away the inhabitants of our sea-girt isle from maritime pursuits, the reckless destruction or partial development of any department of our fisheries cannot be too much deprecated.

But having to deal with the discreditable fact that such destruction has already been carried to a disastrous extent, we are necessitated to inquire whether science indicates the means of speedily repairing such wanton mischief. The Government of France, in the interest of the national economy, has sedulously applied itself to the solution of this important question, and in M. Coste has found at once the intelligent interpreter of its wishes, and the zealous agent in carrying out its designs, or rather his own—for this distinguished naturalist seems to find the Imperial Government constantly anxious to adopt his piscicultural projects.

We have described the singular fishing community of Comacchio. Being desirous now to initiate our readers into the mysteries of *ostreo-culture*, we pray them to accompany us to the shores of the lovely Bay of Naples, near which are the lakes Lucrinus and Avernus, the latter of which, in ancient mythology, was the entrance to hell. About the seventh century the infernal deities were dispossessed of their accustomed haunts, and fled before advancing civilisation. The shaggy woods which cast their shadows on these dismal lakes were felled; a subterranean road (the Grotto of the Sybil) was constructed between Lake Avernus and the town of Cumæ, and piles of splendid buildings replaced these sombre thickets. Baiæ, with its delicious climate and azure sea, and warm baths of various mineral ingredients, became the favourite resort of the luxurious nobles of Rome. Among those who contributed to the gratification of their fastidious appetites was Sergius Orata, a wealthy man, of good parts and popular manners, who set himself to the work of making oysters famous; and this, as Pliny tells us, not from mere love of guzzling, but from greed, " *nec gulæ causâ, sed avaritiæ, magna vectigalia tali ex ingenio suo percipiens.*"—(*Hist. Nat.* l. ix. c. 54.) He organised oyster-parks, stocked with oysters from Brindisium, and persuaded everybody that those he reared in Lake Lucrinus were superior to those of Lake Avernus, or even of the most celebrated places. An arch-epicure doubtless he was, seeing that Cicero (*De Fin.* l. ii.) styles him *luxuriorum magister*, and sorely infected with the love of filthy lucre likewise, seeing that his oyster greed made him, in the interest of his darling molluscs, seize upon public property, which the law could hardly free from his grasp, to the amusement of the Roman wits, who prophesied that, if debarred ostreo-culture in Lake Lucrinus, he would to a certainty carry it up to the house-roofs. Still his guzzling and his greed should not hinder us regarding Sergius as a useful citizen, seeing that he originated a new species of industry, whose usages are still followed for miles round the locality where he pursued it. The scene of this singular industry is nowadays termed Lake Fusaro (between Cumæ and Cape Misenum), a mud-bottomed, volcanic, black, salt lake—the veritable Acheron of Virgil, in fact. The whole vicinity has, from an unknown period, been occupied by spaces, generally circular, filled with stones transported thither. These stones are imitations of rocks, which are covered with oysters from Tarentum, so that each of them forms an artificial bank. Round each of these artificial rocks, generally of the diameter of from six to nine feet, stakes are fastened so near each other as to enclose the central space where the oysters are. These stakes are a little above the surface of the water, so that they can be readily laid hold of and removed when this is desirable. There are also other stakes arranged in long rows, and bound together by a cord, by which are suspended small twigs destined to increase the number of the movable pieces waiting the gathering season.

The oysters generally spawn from June to the end of September, but do not leave their *ova*, like many other marine creatures. They incubate them in the folds of their coverlet (mantle) and among the laminæ of the branchiæ (lungs). There they remain surrounded by mucous matter necessary to their being developed, and within which they pass through the embryo state. The mass of ova, in consistence and colour, resembles thick cream, and breeding oysters are therefore termed milky. The pale tint which first characterises them, gradually, in the process of development, changes to bright yellow, then to a darker yellow, and ends in a greyish-brown or a very marked greyish-violet. The whole mass, which meanwhile is losing its fluidity, probably by absorption of the mucous matter enveloping the ova, is then like a piece of compact mud. This indicates the near termination of the development, and the expulsion of the embryos, and their independent existence; for by this time they can live well enough without the protection of the maternal organs. On leaving the mother they are furnished with a swimming apparatus (its singular nature has been discovered and described by M. le Docteur Davaine), enabling them to move to a distance in search of solid bodies to which they may attach themselves.

The oyster produces not less than from one to two millions of young, so that the animated matter escaping from all the adults in a breeding bank is like a thick mist dispersing from the central spot from which it emanates, and so scattered by the waves that only an imperceptible portion remains near the parent stock. All the rest is dissipated; and if these myriads of wandering animalcules, borne about by the waves, do not meet with solid bodies to which they may attach themselves, their destruction is certain; for those which do not become the prey of the lower creatures living on the *infusoriæ*, fall at last into some place unsuitable to their ultimate development, and are frequently smothered in the mud.

It is thus of the greatest importance that art shall come to the assistance of nature, if man is to be a sharer in the exuberant abundance which is such a striking characteristic of so many species of fish. Millions of embryo molluscs, floating in the ocean at the guidance of the waves, afford the repast provided for certain species of creatures. If man is to partake in this feast, which doubtless is the intention of the beneficent Creator who has given him "dominion over the fish of the sea"—"this great and wide sea, wherein are things creeping innumerable,"—he must exercise intelligence and put forth industry. The waves which surround his dwelling will not, in their random movements, waft these edibles into his hands. In this, as in many other instances, the growing wants of the human race can be adequately supplied only by the union of science with rightly-directed labour. If we wish to have millions of oysters for our million-peopled cities, we must look for the coveted supply not from the rude hands of unskilled fishermen, often so greedy as to senselessly ruin their fields of industry, but from the hands of the scientific naturalist acquainted with the laws of generation peculiar to fishes, and able to turn them to account by the resources of patient ingenuity, which teaches him to wait and reap many harvests from the domain of the waters, rather than impair its future productiveness by reaping a few, in which the immature portion of the crop infinitely preponderates over that which has attained maturity.

This is understood and practised at Fusaro. These stakes and enclosures, which we have described, are arranged for the purpose of arresting this generative dust, and supplying it with points of attachment, just as a swarm of bees settles in the bushes they meet with on their exit from the hive. It does, in fact, become fixed; and each of the animated particles of which it consists grows so rapidly that in two or three years it becomes edible. M. Coste declares that he saw stakes pulled up from the artificial banks, and covered with three distinct crops of oysters, which had been fixed about thirty months. The first of these was fit for the market; while the last, said to be thirty or forty days old, was about the size of a large lentil; "a growth sufficiently surprising," observes M. Coste, "if we remember that, at the time of their expulsion, they were only of the diameter of the fifth of a millimetre" (0.03937 inch). When the fishing season has arrived the stakes and branches are pulled up, and one by one relieved of all the oysters reckoned marketable; and then, after the fruit of these artificial grapes has been gathered, the apparatus is replaced till a new generation yields another crop. At other times the stakes are not removed, and the oysters are merely detached by means of a hook with many branches. The source of these generations remains permanent, perpetuating and renewing itself incessantly by the annual addition of the very few which do not desert their birthplace. The ex-King of Naples, not being remarkable for energy, was too careless to develop this singular species of industry; but what oyster-rearing at Fusaro might yield, in the hands of a private speculator, may be guessed from the fact of its restricted application yielding a revenue of 32,000 francs. M. Coste justly observes: "Transferred to the salt ponds on our coast, it would be a source of wealth to our population; extended and modified in its application to natural banks in the midst of the ocean, it would attain the proportions of an enterprise of general utility. Comparing the mode followed at Fusaro with that followed on the natural banks in the ocean, it is easy to perceive that, if the latter be not suppressed, the source of production must infallibly be soon exhausted. Speculation, in fact, regardless of future generations, which it would nevertheless be so profitable to retain and preserve, only thinks of perfecting the instruments which it uses in tearing from the surface of the beds where they are deposited the oysters which are brought to our markets. Its talent is only applied in rendering the means of destruction more effective; for these beds are precisely those where grow those young ones which, at their birth, have not left the natal spot. Now, as with equal power of destruction it attacks both young and old, it follows that any bed whatever must disappear, simply in consequence of being fished; whereas harvests incomparably more abundant may be taken from it without ever touching the source of production. To attain this important result, it is only necessary to modify the processes so successfully followed at Fusaro. Timber-work, loaded with stones at the base, might be made of many pieces, covered with stakes firmly attached, and armed with iron cramps, &c. Then, at the spawning season, these apparatus could be let down into the sea, either upon or around the oyster-beds; they might be left there till the reproductive seed had covered the different pieces; and cables, indicated on the surface by a buoy, might permit them to be drawn up when it was judged convenient."

All this reads very plausibly, some may think; but will oysters stick to stakes, as, according to theory, they ought? Will old Father Neptune, indignant at the sea being made an artificial fish-pond, not discomfit M. Coste, and make wild work alike with his theories and his apparatus? Your lazy people, who constantly deny that a thing *can* be done till it *is* done before those dull eyes of theirs, which cannot see through the mist of ancient prejudices, laughed at the procreative processes of the philosopher, and, like the Tay fishermen as to the artificial rearing of salmon, were of opinion that "the thing culdna be *just*—for the auld way's best."

It is said that most men of forty are hardly capable

of taking up any new idea, the carrying out of which requires energy and self-decision. Fortunately for the interests of pisciculture at least, the Emperor of the French, though beyond this much-to-be-dreaded epoch in individual history, is remarkably receptive of new suggestions, so that novel implements of war for slaying men, and of industry for multiplying oysters and other fish, interest his busy intellect. Hence he is a keen pisciculturist. His imperial will, directing the national wealth, brooks no long delay between the announcement of a theoretical improvement and the carrying of it out in practice. M. Coste having roused his attention to the importance of pisciculture in general, and of ostreo-culture especially as the means of replenishing the exhausted oyster-beds on the coasts of France, measures were speedily adopted with the view of experimentally testing the value of M. Coste's suggestions. The result is before us in a report to the Emperor, with a copy of which M. Coste has obligingly favoured us. We shall avail ourselves of its contents, in the hope that their publication in this country may stimulate to similar modes of procedure those who lament the rapidly-advancing destruction of the oyster-beds on the shores of Great Britain and Ireland.

The philosophic naturalist commences with the assertion that the sea can be cultivated like the land; and that to the State belongs the duty of putting into exercise those appliances of which science guarantees the success, and of then handing over to a grateful population the harvests prepared by its care. He deplores the destruction of eighteen out of the twenty-three formerly valuable oyster-banks at Rochelle, Marennes, Rochefort, &c., the oyster-rearers at which places, being no longer able to stock their ponds, have to go to the British coast for a supply, which they find both expensive and insufficient. The fifteen oyster-banks in the bay of Saint Brieuc, which formerly yielded an annual revenue of about 4000 francs, and employment to 1400 men, who manned 200 boats, are now reduced to three, employing 20 boats. And with this diminished production there is a largely increased demand, railways having multiplied the means of intercourse with the coast and the numbers of the population anxious to share in the products of the ocean. M. Coste consoles the Emperor by the assurance that for this deplorable state of things there is a remedy, easy of application, and certain to augment immensely the means of public alimentation. He proposes, in short, to put in practice, on a large scale, the mode of rearing oysters which he had witnessed at Fusaro; and suggests the bay of Saint Brieuc as the locality of his interesting experiment. The problem for solution is this: Of the from one to two millions of young produced by an oyster, only from ten to twelve remain attached to the shells of the mother. How is this to be prevented, so that these swarming molluscs shall be fixed on the bottom of the sea, instead of being scattered by its waves? The method suggested does not deprive the natural beds of any portion which they generally retain, and yet it may yield to us incalculable riches. In order to obtain them, we have only to deposit in sheltered banks, hurdles, and stakes still retaining their bark, and kept at the bottom by weights, and laid flat, so as not to interfere with navigation. The progeny of the oysters placed below these will rise like a cloud of animated dust through the branches, and the embryos of which it consists will encrust every part of the apparatus devised for the reception of the seed. After remaining on it for a certain period, the young shell-fish will spontaneously be detached, and fall to the bottom (previously cleaned by the drag), like the seed of the sower in the ground prepared by the harrow. And then, in a passage of seductive eloquence, the enthusiastic naturalist promises that, if his imperial patron will supply the very moderate expenditure to be incurred, the whole coast of France shall be converted into a long chain of oyster-banks, interrupted only in those places where there is an accumulation of mud. Oysters shall swarm on every rock from Dieppe to Havre, from Havre to Cherbourg, from Cherbourg to the depopulated banks of La Rochelle; and other famed localities shall resume more than their ancient prosperity.

Chimerical rashness! cried the obstructives, who are constantly frightened by the noise of their own nonsense, which they mistake for that of lions in the way. "We shall see," thought the Emperor, "when M. Coste has employed the means which I now place at his disposal."

Such is the substance of Report the first. In Report the second we advance from the promise to the performance; from the clamours of the ignorant to the felicitations of the wise, congratulating a learned colleague on the happy issue of an experiment of great public utility. It is worth while to relate the simple but ingenious methods employed in restocking the oyster-depopulated bay of Saint Brieuc. The immersion of the oysters on ten previously designated sites began in March and ended in April. Within this brief period thirty millions were deposited in different parts of the bay, representing the space of a thousand hectares—one hectare is equal to 195 square perches. In order that the operation might be conducted with the regularity of an agricultural process, and that the oysters might be deposited regularly, and at such distances as not to interfere with each other, a steamer was employed to tow a fleet of boats laden with oysters in baskets, which were emptied in the spaces marked by buoys. An ingenious device was fallen upon to induce the oysters immediately to fasten themselves in their new locality. Shells of oysters and other shellfish were collected from different quarters, and thrown down upon the banks to be operated upon. This rubbish, the removal of which every year used to be expensive, was converted into points of attachment, on which the newly transported oysters speedily fastened, and began the work of reproduction immediately, owing to their being on the point of spawning. Every embryo was thus provided with a solid body on which to fasten. But the difficulty of arresting the seminal fluid of the procreant oysters had also to be overcome. Long lines of hurdles, made of twigs of from four to five metres, were ranged across the banks, and were kept floating above the gravid oysters at the distance of from thirty to forty centimetres, being kept in their places by being suspended from their centres by a rope fastened to ballast-stones. These were laid down by men, protected by a breathing apparatus, and instructed to deposit around them a certain number of parturient oysters. The ropes were found to rot quickly; and M. Coste now recommends the substitution of chains of galvanised iron. Hardly six months had elapsed when the promises of science were astonishingly verified. The result surpassed its most sanguine hopes. The breeding oysters, the shells which covered the bottom, the very strand even, were ascertained to be covered with oysterlings! "Never," exclaims M. Coste, "did Cancale and Granville, at the time of their highest prosperity, exhibit such a spectacle of productiveness. Every part of the hurdles is loaded with clusters of oysters in such profusion as to resemble the trees of our orchards, when in spring their branches are covered with a profusion of blossom. They should be termed actual petrifactions. Seeing is necessary to believing such a wonder. I have sent to your Majesty one of these apparatus for collecting seed, in order that, with your own eyes, you may judge of the riches of these hurdles. The young oysters which cover them are already of the size of from two to three centimetres. These, then, are fruits which only require eighteen months in order to ripen into an immense harvest. There are even twenty thousand in a single hurdle, which occupies not more space in the water than is occupied by a stalk of corn in a field.

Now, twenty thousand edible oysters are worth 400 francs; their current price, when bought on the spot, being 20 francs per thousand. The revenue from this species of industry will consequently be inexhaustible; for we can submerge as many seed-collecting apparatus as we choose, and each adult, forming part of the deposit, produces at least two or three millions of embryos. The bay of Saint Brieuc will thus become an actual granary of abundance, if, by the union of the banks already created, we convert the whole of it into a vast field of production." The experiment made in the bay of Saint Brieuc is too striking to allow us to be blind to the light of its instruction. It proves, by a brilliant result, that wherever the bottom is free of mud, industry, guided by science, can create in the bosom of oceans, fertilised by its care, harvests more abundant than are yielded by the land.

This most interesting Report concludes with an earnest appeal that the Emperor would be pleased to order the construction along the coast of various scientific establishments for the permanent study of facts connected with ichthyology, and for the further elucidation of its mysteries. " In an age," observes M. Coste, "in which, by an exalted application of the laws of physics, an invisible flame conveys thought along the conducting wires with which human genius entwines the globe, physiology will exercise its empire over organic nature by an application of the laws of life."

These reports of M. Coste, so suggestive of physiological problems of the highest importance, also contain practical suggestions of great value for the improvement of oyster-fishings. He insists that these shall be made subject to regulation both as to the time during which they shall be permitted, and the manner in which they shall be conducted. He protests against beginning oyster-fishing in September, and recommends that this be delayed till February or March; because, he argues, though in September spawning be over, the shells are covered with a recently-formed population: so that if, at the very time when this repeopling has been accomplished, the fishing be commenced, the damage will be almost as great as if it had begun during the period of gestation. Along with the adult oysters we take the young ones with which they are encrusted. The drag thus makes havoc of fields in full germination, like a rake drawn across a tree in full blossom. By delaying the opening of the fishing season till March, the greater part of the young oysters will be detached, and those which still adhere can be removed, and either restored to their position, as the law prescribes, or preserved in rearing-ponds, as at Cancales. There is no force in the objection that three months' fishing will not suffice to supply the markets, seeing that in May oysters begin to be *milky*, and their capture is then prohibited. The oysters consumed during the greater part of the year are not taken from the sea, but from the various "parcs" and "claires" (reservoirs), in which for several months they are prepared for the market. A curious account is given of the "claires" of Marennes, in which the fishermen deposit oysters in order that they may become green—a result, it seems, which greatly improves their flavour and tenderness. These "claires" are, as it were, so many inundated fields established here and there on both sides of the creek of La Soudre. They differ from ordinary ponds and parks in not being, like these, submerged at every tide, but only at spring-tides; a more frequent submergence would frustrate the desired end. Irregular in their dimensions, they generally contain about three hundred square metres of superficies. They are filled at spring-tides, and the water is permitted to remain till the soil, thoroughly impregnated with saline particles, resembles that of the bottom of the sea. They are then emptied, and the soil is *dressed*—dried, that is, and levelled like a garden walk or barn floor; all foreign substances, such as plants dead or living, carefully removed, in order that in this smooth surface, hardened by the rays of the sun, nothing may obstruct the free development of the edible mollusc which is there to be reared. In September, men, women, and children are all busy collecting the oysters to be deposited within them; and it is a curious fact that only the young oysters taken from this locality can be made to acquire the good qualities resulting from the process of greening. Old ones, or those from other places, acquire the greenness; but a knowing palate detects in them a certain bitterness. Hence the rearers who seek to please their customers are very careful only to admit young "natives" into their reservoirs. An oyster deposited in the "claires," at the age of from six to eight months, must remain there for two years before attaining the proper size; and three or even four years are requisite in order to reach the perfection of Marennes oysterhood. A long delay this, no doubt, but abundantly rewarded, it appears, seeing that the beds of Marennes furnish annually five millions of oysters, valued at £83,334. As is not uncommon elsewhere, the laborious task of disposing of the oysters in the neighbouring towns falls to the women, whose acquaintance with city life gives them assurance and a knowledge of the fashions, which is exhibited on Sundays with such a transmogrifying effect that, in damsels of flexible shape, coquettish air, and easy walk, it is not easy to recognise the fishwives whom yesterday we met in the town, selling oysters before our hotel, or at the corner of the streets.

In Scotland oysters are nowhere subjected to the treatment called "greening," by which their flavour and tenderness are greatly improved. The cause of the *viridity* of the oyster is disputed, some ascribing it to an insect, others to disease of the liver, and others to the nature of the soil in the "parcs" or "claires" in which they are deposited. The analyses of M. Berthelot and others favour the last supposition, and seem to prove that the greening property is due to a slight excess in the amount of the sulphate of iron and the chloride of sodium. The introduction into Scotland of the greening system, and its further extension in England, are desirable, as the means of insuring to our markets a more regular supply of improved oysters. If at Marennes the viridification of oysters yields annually five millions so improved as to be worth £83,334, where is our Scotch "*canniness*," when it is notorious that, though tempted by the most favourable circumstances, we have never even tried to green an oyster! Nay, it is asserted, we are in such a truly primitive condition in regard to the best of our molluscs that our oysters are all eaten *puris naturalibus*, and without any scientific treatment whatever.

Large quantities are sent from Prestonpans to fatten in bays near the mouths of the Thames and the Medway. Scotch oysters, like Scottish maidens, are, it seems, attracted to the metropolis for the perfection of their education. The pity is that, unlike young ladies, they do not return to gladden Scottish eyes by the display of their fully-developed virtues. As advocates of home-training, we object to the "natives" of Scotland being domiciled in English "parcs" for the sole use and comfort of English epicures. As we cannot hope to make acquaintance with them by their being sent back to us, why not rear them at home according to the most approved methods, to the mutual satisfaction of eaters and venders, their improved qualities commending them to the palates of the one while filling the purses of the other? Why should any branch of our native industry be left undeveloped? When storms keep the fisherman unwillingly on land, and so long occasionally that he is equally ill off for something to do and something to eat, why not find a remedy for this misfortune in the semi-aquatic and very profitable employment of rearing oysters? As they thrive best in estuaries and bays, there is no lack of suitable localities. And as the estuary of the Clyde is oysterless, we commend to some energetic fishmonger the enterprise of supplying Glasgow with "the food that feeds, the living luxury," by imitating the processes by means of which their ancient productiveness has been more than restored to the exhausted oyster-banks along the coasts of France.

When the Clyde is no longer oysterless, let us hope that the motto of the western metropolis of Scotland —"Let Glasgow Flourish"—shall, in addition to its device of a salmon, exhibit the figure of an oyster; in token of benefit received from the successful prosecution of oyster-culture.

MARITIME PISCICULTURE—MUSSEL CULTURE.*

We forget which of these poetical heathens, Horace or Virgil, was so shockingly ignorant of the true nature of things as to call the ocean "*dissociabilis.*" All the world now knows that the sea is the highway of nations; and, more than that, poetry nowadays tells us much of the treasures of the deep, where

> "Life, in rare and beautiful forms,
> Is sporting amid the bowers of stone,
> Safe from his rage, when the god of the storms
> Has made the tops of the waves his own."

Better still, a taste for natural history is now widely diffused, and the modern invention of domestic *aquaria* has made us acquainted with many of the most beautiful of marine productions, animate and inanimate. The ladies, in particular, enjoy the opening up of this, by most of them, heretofore unread page of the mighty volume which reveals the Creator's glorious attributes. We know of one household in which a most lively interest is felt in the singular process by which provision

* Voyage d'Exploration sur le Littoral de la France et de l'Italie. Rapport à M. le Ministre de l'Agriculture, du Commerce, et des Travaux Publics, sur les Industries de Comacchio, du Lac Fusaro, de Marennes, et de l'Anse de l'Aiguillon. Par M. Coste, Membre de l'Institut, Professeur au Collége de France. Paris : Imprimerie Imperiale.

is made for the casting of the shell of the *crab*. A very domestic shell-fish is our friend, and so sociable in his ways that we cannot ascribe the origin of the term "*crabbed*" to the known surliness of his species. He walks along the carpet molesting no one, and at four o'clock, when the family meal is served, never fails to ask leave to share in the repast by vigorous taps on the side of the aquarium.

Philosophers, moreover, and political economists, puzzled with the problem of how to provide food commensurate with the increase of the world's population, are seriously trying to convince us that the ocean is rich in substances more precious than coral, pearls, and amber; that it is, in fact, the great storehouse within which the Universal Parent has laid up exhaustless supplies of food, waiting to be drawn forth by human labour and intelligence. It is indeed most marvellous that multitudes should be pining with hunger in a land like ours, literally set in an ocean of plenty. We may be unable to indulge in the luxury of *hippophagy*, horse-flesh being costly; but with the sea as our fish-pond, and pisciculture capable of endless development alike in salt water and in fresh, only the laziest or stupidest of men, or the sick and the helpless, should be heard bemoaning themselves that they perish with hunger. Truly in our Father's house there is food and to spare. The green crested waves, glancing in the breeze, are as suggestive of plenty as are the harvest-fields loaded with the products of the land. Nay, agriculture yields no such abundant and inexpensive harvests as may be reaped in rivers and oceans teeming with animal life, rich in the choicest elements of nutrition.

> "Oh, what an endless work have I in hand,
> To count the Sea's abundant progeny!
> Whose fruitfulle seede farre passeth those inland,
> And also those which wonne in the azure sky;
> For much more eath to tell the starres on hy,
> Albe they endless seem in estimation,
> Then to recount the Sea's posterity,

So fertile be the flouds in generation,
So huge their numbers, and so numberless their nation."
—SPENSER.

To demonstrate this has been the object of our articles on Pisciculture.

Our present article relates to the rearing of a shellfish, by no means so valuable as the oyster, and yet deserving of cultivation not merely as an article of human food, but also as the bait largely used in the capture of valuable species of sea-fish. We refer to the mussel, not to be always eaten with impunity by human beings—for on some constitutions it acts perniciously, and sometimes fatally, when it has been taken from filthy harbours, or from copper-bottomed ships; but valuable, notwithstanding, as the bait largely used by our fishermen. The procuring of it is the laborious and degrading work of the fisherwomen—physically injurious, because pursued often in the most inclement weather, and at the earliest dawn, or late at night, as necessitated by the state of the tide; and mentally hurtful likewise. The fisherwomen on the east coast of Scotland are deplorably ignorant, being hard-wrought in gathering bait and fastening it to the hooks; and this ceaseless bait-gathering is found to be a serious difficulty in getting them to give anything like regular attendance at adult classes. Moreover, the mussel in certain localities is becoming exceedingly scarce. In the vicinity of Arbroath, for example, it is procurable with such difficulty that the fishermen have long been in the habit of seeking a supply from the opposite coast of Fife, at the mouth of the Eden, near St Andrews. *There*, too, the supply has failed them from some unascertained cause; and the curious result is this: The estuary of the Clyde supplies the estuary of the Tay with mussels for bait; the expensive means of transport being the railway.*

* A gentleman acquainted with the fishing villages round Arbroath has obligingly supplied the following information:—

Now, for this state of matters—so directly injurious to the hardy race of fishermen, their wives and daughters, and indirectly to the public by raising the price of fish—there is a remedy. Pisciculture, so rich in promise to the epicure fond of luscious oysters, comes to the aid of the humble fisherman likewise, and by multiplying mussels will ease the back of many a "Maggie Mucklebacket," so far as the gathering of bait is concerned, though doubtless, to the mutual benefit of all parties, it will enable her amphibious husband to prosecute his calling more regularly, and thus be the means of throwing a heavier load on her willing shoulders, in the form of "creels" brimming with "caller haddies."

We trust that those interested in the prosperity of our fishing villages will have the benevolence to tell their inhabitants that their women need not wretchedly toil for bait, and that the gains of the men should no longer be diminished by the purchase at a distance of what may inexpensively be multiplied indefinitely at their doors. Mussels, in short, as food for those who like them—be they men or fishes—may be made to propagate where we can lay our hands upon them whenever we choose; so that fishwives need no longer be bedraggled, toil-worn gatherers of this kind of bait, nor need the fishermen of our eastern coast be indebted to those of the western for the means of pursuing their piscatory labours.

Mussels seem to be more in repute in France than in

"The whole of the mussels used for bait in this quarter is brought from the Clyde. The price of the bushel-basket was formerly 1s. per basket at the Eden, and this has now been raised to 1s. 9d. per basket. The cost to each man per annum for bait was formerly about £4.

"The mussels found in the Clyde are substantially free to every one. The dredging for them costs about 10s. 6d. per ton; the carriage per rail to Arbroath 15s. per ton, or about 1s. 2d. per bushel-basket. In a ton of mussels there are about 24 bushel-baskets. A boat could bring from the Eden from 60 to 80 baskets.

"The Clyde mussels are now sent as far as the Moray Firth!"

this country. M. Coste speaks of the beautiful mussels daily used at table, and inform those who use them that they mistake in supposing that, like oysters, they come from natural banks, where they live in their wild state. They are not aware, it seems, of the artifice by means of which human industry gives such an improved shape and flavour to the lean, little, bitter, and often unwholesome mussel, so numerous among the rocks and sands of the French coast.

The mussel has not been much written about. The most curious details regarding mussel-rearing are found in a very rare book, published at Rouen in 1598, and bearing this title: 'Théâtre des merveilles de l'industrie humaine.' It is carried on in "*bouchots*," a term, according to M. Coste, contracted from *bout-choat*, which is a compound from the Celtic and the Irish—*bout*, an enclosure, and *choat*, wooden. The latest information regarding this species of industry appears in the 'Annals of the Agricultural Society of Rochelle for 1846.'

An interesting story is connected with this importation of an Irish word into the dialect of the western coast of France. In the bay of Aiguillon, not far from Rochelle, a poor shipwrecked Irishman originated an industry which has lasted more than six hundred years, and been the means of furnishing a comfortable livelihood to three thousand people. In the year 1235 a bark, laden with sheep, and manned by three men, was driven from the Irish coast by a furious gale, and dashed among the rocks near the harbour of Esnandes. All would have perished but for the gallantry of the fishermen, who hastened to help the ship in distress. Their utmost exertions could only rescue one of the crew—the skipper, Walton—the founder of the first *bouchot*, a beneficent invention which has for long enriched one French province, and the application of which to other coasts will, M. Coste predicts, one day cause the yet obscure name of the inventor to be enrolled among the most useful benefactors of humanity.

Accident led Walton to his happy discovery. Compelled to live by his wits, he set about exploring the miserable locality on which he had been cast, in order, if possible, to discover the means of subsistence. This he found in the netting of sea-birds in the bay of Aiguillon. He was not long in observing that the mussels attached to the stakes supporting his nets were larger and better-flavoured than those living in their natural state in the mud. Profiting by the hint, he planted more stakes, which in turn were covered with mussels. Having thus had proof that it was possible to practise mussel-rearing so as to assume the proportions of an extensive cultivation, he set about the construction of permanent erections for the reception of the young of the mussel. These consisted of two angular lines of stakes, with the apex towards the sea, and distant from each so as to form nearly an angle of 45°. Along each of these lines, at the distance of two or three feet from each other, he fixed strong stakes of from ten to twelve feet high, which, to the depth of six feet, were driven into the sand, the intervening spaces being interlaced with hurdles and branches capable of resisting the action of the waves. At the outer extremity of the angles he left a space of three or four feet, for the reception of an apparatus intended to capture the fish passing through these double lines, with the flow of the tide—thus ingeniously combining fishing with mussel-rearing. These double lines were so arranged that the juxtaposition of two sets of them formed the letter W; and M. Coste fancies that, by thus inscribing on the sand the initial letter of his name, Walton was preferring his claim to that public gratitude which he doubtless expected. Certain it is that this is still the form of the five hundred *bouchots* which now cover half the bay of Aiguillon.

He must have been an ingenious man, as is shown by his admirable device for enabling him to explore the bay at low water, and construct his apparatus. Sailing or walking along the muddy and yielding surface being impracticable, he contrived a "*poussepied*," a kind of boat impelled by the foot in this fashion:—It is a simple wooden box, nine feet long and eighteen inches wide and deep, with one extremity curved into the shape of a prow. The *boucholeur*, or mussel-man, as he may be termed, places himself at the hinder end, with his right knee on the bottom, and, leaning forward, grasps the sides with both hands, while the left leg, which is outside, and protected by a strong boot, does the work of an oar, by being alternately struck into and withdrawn from the sand on which he rests. "The return of a fleet of 160 such singular vessels is a curious sight. Emerging at all points out of the forest of stakes, they are like a flock of birds chased by the tide. Without having seen them, it is impossible to form a conception of the grotesque manœuvres of the strange squadron." By joining together three of these boats, the one in the middle being loaded, and the two on either side of it being leg-driven, as we have explained, large quantities of stakes and hurdles can with ease be transported to their place of destination.

We indicate this novel form of vessel, as fitted to be useful on our coasts when the apparatus of the fishermen cannot be approached by the common cobles requiring a considerable depth of water.

At a certain season of the year the manœuvres of these boats would be very difficult, but for the operations of a small crustacean (*Corophium longicornis*), which, in its chase after the sea-worms on which it feeds, levels the deep furrows and temporary inequalities of the sand, which, being hardened by the sun, impedes the movements of the *boucholeurs;* and thus, in a few weeks, these little creatures do what thousands of men could not do in a whole summer.

Walton never gave up the use of stakes without hurdles, but, fixing numbers of these along the shore, made them serve for the filling up of gaps in the hurdles not fully occupied by the spawn for the season; and the very next spring the beautiful mussels reared in these artificial *parks* were preferred in all the markets. His neighbours, struck with the profit derived from his industry, imitated his example with such eagerness that the whole sands were speedily covered with *bouchots;* "and," observes M. Coste, "when writing these lines a forest of two hundred and thirty thousand stakes is employed for the permanent support of a hundred and twenty-five thousand hurdles, groaning under the annual harvest, which could not be stowed away in a squadron of ships of the line."

This immense rearing of mussels, a species of food little esteemed among us, is deserving of consideration, because opening up new fields of industry, and furnishing the labouring classes with an alimentary substance of no mean value. Our mussels neither enter largely into the national subsistence, nor ought they to do so in the native state in which alone we are acquainted with them. But improved by the processes pursued on the western coast of France, it cannot be questioned that mussel-rearing would be profitable, because the public would soon learn to appreciate their taste and appearance. And when our readers peruse our subsequent statements as to the economical results of Walton's process, it must be borne in mind that we speak of mussels judiciously manipulated *ab ovo* until they grace the dinner-table of people quite able to appreciate culinary dainties. If the French eat millions of mussels, we may depend upon it that they are worth eating, and that the sooner we make acquaintance with such educated molluscs, the better will it be for the interests of our fishermen and of the public stomach. Between a French-bred mussel and one picked up by chance at the mouth of the Tay or the Clyde, there is such a difference as at once to explain the comparative value of mussels in France and Great Britain. The *bouchots*, the employment of which occasions this difference, consist of four parts, distinguished by different names, according to their distance from the shore. Those farthest out are simple stakes, never uncovered except at neap-tides. These solitary stakes are the best situated for the preservation of the young mussels which fasten on them. Everywhere else these excessively delicate creatures would be too often dry, and could not withstand the prolonged heat of the sun nor severe cold. These stakes are therefore the special points where are left to accumulate the whole of the young ones, which are afterwards to be transplanted to the empty or partially-covered hurdles, which the sea leaves dry more frequently. We may here remark that the different operations are designated by agricultural terms, such as sowing, planting, transplanting, weeding, pricking-out, and reaping.

Towards the month of April the *seed*, fixed in February and March to the single stakes, is scarcely the size of lintseed; in May it reaches the size of a lentil; in July that of a French bean, and at this stage is transplanted. At this season the mussel-gatherers detach the young from the stakes by means of a rake with a handle, and proceed to fasten them, one bunch after another (each wrapped in a piece of old net), among the branches of the hurdles; care being taken, however, that each separate bunch shall be at such a distance from the next one, that when grown they shall not mutually incommode each other. The net soon rots, and thus presents no obstacle to their expansion. In the end they have grown so much as to touch each other; so that these immense palisades, when the fully-developed bunches meet together, resemble the sides of walls blackened by fire.

Arrived at this stage, they can be weeded in order to afford room for younger generations, and be pricked-out—an operation which is effected in the manner already described; that is, by wrapping each bunch in a piece of net before consigning them to their new habitat.

At the end of ten or twelve months the mussels are marketable. They now undergo their last transportation to the in-shore hurdles, where they are at hand, and do not suffer though the sea leave them dry twice every day. It is a singular fact that mussels growing together on the same hurdles do not all possess the same qualities. Those on the highest rows are better-flavoured than those on the middle rows, and the latter again are deemed inferior to those lower down, which are covered with mud whenever the waves agitate the bottom. But M. Coste affirms that "even the least esteemed of these mussels are much preferable to the finest gathered in the sea."

Mussels, being so abundant and cheap as to have become the daily food of the labouring classes, are sold during the whole year; but from July to January they are in the greatest perfection. From the end of February to the end of April they are *milky*, and, like oysters at the spawning-time, lose the qualities they previously possessed.

Having thus explained how mussels become an edible of value, we trust that we have indicated to our hardy fishermen a source of profitable employment, as yet only partially developed. Why should mussels in driblets only—bitter and tough, moreover, and fit only for bait—be all that we can exhibit on our Scottish coasts? Reared by millions, and improved by cultivation, they support in comfort several thousand French fishermen, the owners of a hundred and forty horses and ninety carts employed in supplying with this—by us neglected—shellfish the towns of Rochelle, Rochefort, Poitiers, Tours, Saumur, &c.

Each *bouchot* bears a harvest of from a hundred and twenty to a hundred and fifty pounds' weight, worth from £84 to £105. M. Coste notes the important fact that, since this industry was largely prosecuted, there is in the locality enriched by Walton's discovery no healthy man who is poor. Those whom infirmity condemns to enforced idleness are generously supported by their neighbours. Such unfortunates regularly wait the arrival of the boats, and receive one handful of mussels and another of small fish, accompanied with kind looks and friendly words. Every housewife, moreover, when baking, puts aside a piece of dough for the disabled; and the bakers, collecting all the pieces, make them into loaves, which are baked *gratis*. M. Orbigny, in his report to Government on the inhabitants of the maritime communes of the bay of Aiguillon, likens them to the settlements of the Moravians in North America and Germany. Their characteristics, he declares, are industry, good morals, gaiety, and happiness; little quarrelling or drunkenness, cheerful homes, and a religious love of hospitality and truthfulness.

We appear to have got into the millennium! and all this from the rearing of mussels! We hope some benevolent soul will reprint this in big type for the information of the fishermen of Broughty-Ferry, Carnoustie, Arbroath, Auchmithie, where dwell the Cargills, the Smiths, and the Swankies, of whose pre-eminence in the virtues chronicled by M. Orbigny we have yet to learn.

We are convinced that the resources of pisciculture may essentially ameliorate the condition of our British fishermen and their families. We have therefore expiscated for their behoof the valuable information accumulated by M. Coste regarding eels, oysters, and mussels. But it will never be brought under their notice, unless we have excited some intelligent maritime proprietor to take up the matter, and prosecute it with assiduity, as the commencement of a kind of industry sure to be abundantly rewarded.

FISH DIET, AND ITS EFFECTS ON THE HUMAN CONSTITUTION.

TURNING to the article "Aliments" in the 'Edinburgh Encyclopædia,' we read: "From a comparison of their respective qualities and organisation, we might have concluded that fish would, in equal weight, afford a less nourishing aliment than flesh, and of more difficult digestion and assimilation. Experience comes in support of this conclusion. The Roman Catholics who, during the forty days of Lent, rigorously abstain from the use of flesh, but indulge freely in a fish diet, are said to be less nourished by it, and to become sensibly thinner, as Haller indeed tells us he had himself experienced. The general practice of using higher seasonings and sauces with fish, and the custom so common in our own country of taking a dram after this kind of food, show plainly enough what is the general experience of mankind with regard to the alimentary properties of fish."

We demur as to the logic of the Encyclopædist. Because some people use horse-radish along with roast-beef, we humbly think that this is not satisfactory evidence as to the innutritious qualities of beef, or as to its being of difficult digestion. And because, in Scotland, fish is facetiously known as "the Latin for a dram," we are not prepared to jump to the conclusion that, in "the general experience of mankind," fish is found to be hard to digest. We suspect that the use to which the Scotch apply their Latinity is the real though latent cause of the digestive remorse sometimes consequent upon " a fish-dinner;" in short, that the dram is the drawback to the satisfactory " assimilation " of the fish! We have not forgotten the experiment of Dr Beddoes, who found that animals to which spirits had been given along with their food, digested one-half less than other animals from which this stimulus had been withheld. In short, we are not disposed to allow the question as to the alimentary properties of fish to be settled by the limited experience either of Scotch dram-drinkers, or of Roman Catholics putting their constitutions to an unnatural trial, by suddenly, at Lent, beginning to disuse butcher-meat. We think it much more satisfactory to refer to the long experience of a fish-eating community, such as that of Comacchio.

"The inhabitants of Comacchio," observes M. Coste, —"those of them at least connected with the fishery and its associated avocations—in addition to wine, chestnuts, flour-pudding, and some fruits, live upon fish alone, and, above all, upon eels. And yet this diet, far from injuring the public health, maintains it in the most flourishing condition. Persons submitted to the permanent influence of this diet are robust, and live as long as those who, in other countries, live on butcher-meat. Their elevated stature—the breadth of their chest—the muscularity of their limbs—the elasticity of their bodies—their animated look—their bright complexion—their thick black hair—are proofs of vigour as striking as can be seen in any other part of Italy."

The women, too, are remarkable for " their Grecian profiles—their flowing hair—their elegant figures, whose admirable proportions never yield to the uncomeliness of obesity. They are fruitful mothers, who, after paying tribute to the race whose integrity they preserve, often reach decrepitude without any infirmity condemning their hale old age to inactivity." M. Coste is undoubtedly right in maintaining that the experience of this singular community, which for hundreds of years has preserved its vigorous type by means of a diet composed almost exclusively of three kinds of fish, is a memorable example of what Governments may do in providing sustenance for advancing populations, by encouraging the use of a diet which constitutes so very small a proportion of the national food. Comacchio is remarkably healthy. Intermittent fever, so common in the neighbouring marshes, is not frequent, and scurvy is of exceptional occurrence. And thus, when the young of the neighbouring district are of feeble constitution, or threatened with consumption, they are sent to the lagoons of Comacchio, to share in the toils and the fare of the fishermen. In short, the value of a fish diet is demonstrated by an experiment perfectly unique in the history of the world.

In his commendable zeal for the increase of the means of national aliment, M. Coste thinks it necessary to demonstrate that there is no foundation for the popular opinion that fish-eating leads to an embarrassing increase of population. Because fish are amazingly prolific, certain wiseacres appear to have jumped to the conclusion that fish-eating races must be so likewise. This idea prevails at Comacchio, one of the islands of which is famed as the island of fecundity—a fame resting on the fact that a residence thereon made a mother of the previously barren spouse of a nobleman of Ferrara. M. Coste attributes this interesting event to the improved health of the illustrious lady rather than to the prolific virtues of her fish diet, and proceeds, without remorse, to demolish the hopes of childless women, by demonstrating, statistically, that they need not repair to Comacchio, expecting *there* to have their "reproach among women" taken away. The fact is, the increase of population at Comacchio is below that of the inland parts of Italy; the births being in the proportion of 1 in 29 in the one case, and 1 in 27 in the other.

It is curious how long-lived are the fancies of public credulity. The belief in the superior prolific powers of piscivorous people prevailed so far back as the days of Hippocrates. It was not destroyed by the result of the experiments made upon himself by that profound philosopher and celebrated physician, Haller. Finding himself debilitated by living on fish, he draws the conclusion that the Romish Church does well in prescribing the sole (or frequent) use of fish to monks—"*generationi non destinati.*"

We must confess, however, that we are disappointed by our researches as to the effects of fish on the human constitution. The medical oracles speak with a dubiety and obvious ignorance of the whole matter which are unsatisfactory. In the 'Transactions of the Royal Society of Edinburgh' there is an interesting communication by Dr John Davy, read in April 1853. He asks—"What are the nutritive qualities of fish, compared with other kinds of animal food? Do different species of fish differ materially in degree in nutritive power? Have fish, as food, any peculiar or special properties? These are questions, among many others, which may be asked, but which, in the present state of our knowledge, I apprehend it would be difficult to answer in a manner at all satisfactory."

As the doctors confess their incompetency to guide us to positive conclusions regarding such questions, let us be thankful that the discriminating powers of the human stomach have long ago determined that salmon and mackerel are nourishing and delectable, though it was not till the year 1853 that Dr John Davy told the world that the amount of solid contents in 100 parts of beef was 26, while of salmon and mackerel the solid contents of 100 parts amount to 29 and 27 respectively. For the credit of science it is consolatory to know that its conclusions are also in harmony with experience regarding the important questions whether fish as a diet is more conducive to health than the flesh of the animals usually eaten, and especially whether it prevents scrofulous and tubercular disease. " I am disposed to think they are," observes Dr Davy. " It is well known that fishermen and their families, living principally on fish, are commonly healthy, and (may I not say?) above the average; and I think it is pretty certain that they are less subject to the diseases referred to than any other class, without exception. Dr Cookworthy has, at my request, consulted the records of the Public Dispensary at Plymouth; and it appears that of 654 cases of confirmed diseases of the lungs, the probable result of tubercles (234 males, 376 females), the small number of 4 only were of fishermen's families, which is in the ratio of 1 to 163. The entries from which the 654 cases are extracted, Dr Cookworthy states, exceed 20,000. He assures me that, had he taken scrofula in all its forms, the results would, he believes, have been more conclusive."

The returns of the Registrar-General render the fact of this exemption of fishermen still more remarkable, especially considering the prevalency of tubercular consumption in the working-classes generally throughout the kingdom. The cause of this sanatory action of fish is ascribed to the presence of iodine in their composition. " In every instance in which I sought for this substance in sea-fish, I have found distinct traces of it, and also, though less strongly marked, in the migratory fish, but not in fresh-water fish." " That iodine," Dr Davy goes on to remark, " should enter into the composition of sea-fish, is no more, perhaps, than might be expected, considering that it forms a part of so many of the inhabitants of the sea on which fish feed; to mention only what I have ascertained myself in the common shrimp, in an unmistakable manner, and also in the lobster and crab, and likewise in the common cockle, mussel, and oyster."

The medicinal effects of cod-liver oil in mitigating or curing pulmonary affections is well known, and appears also to be due to the presence of iodine, the value of which in the treatment of that shocking malady goitre is likewise acknowledged. The most recent authority on this matter is M. Le Docteur Fonssagrives, the learned author of 'Traité d'Hygiene Alimentaire.' What he says of oysters may perhaps, within certain limits, be applied to all marine shell-fish. "May we," he asks, "not believe that these molluscs, living in a medium rich in iodine, store up this substance, and communicate it to those organisms which they feed, without causing them to incur the least risk of that constitutional iodism which epicures constantly incur with impunity when feasting on the products of Ostend and Cancale? I am in the habit of recommending oysters to weak, lymphatic, soft-fleshed children, and of making them drink a considerable quantity of the liquid which the oyster pours out when opened; and I think experience justifies me in ascribing to this treatment a very favourable effect in various forms of lymphatism."

M. Charles Bretagne, in a communication to the 'Bulletin of the Imperial Society of Acclimatation,' declares that the species of clam scientifically known as *Venus verrucosa*, is "an alimentary pearl," deserving to be extensively used, and requiring protection, because, inhabiting sandy shores, it is easily discovered by children, walkers, bathers, and everybody. He recommends that it be introduced to the shores of the Mediterranean —shores which will soon be frequented by all consumptive patients, who will sometimes find a cure for their maladies, very often a prolongation of life, and, at all events, an easier death.

"Shell-fish," he maintains, "will here be specially appreciated, for, besides being an agreeable food, it is a powerful auxiliary in curing chest-complaints. Doctors are now unanimous in combating poverty of the blood by means of iodine. Men of science desire that their patients shall assimilate their remedies by the most natural process—that is, as food. Everything induces us to follow this rational procedure; and in creating a taste for *Venus verrucosa*, and multiplying it, we provide at once an agreeable aliment, and a fresh and salutary medicament."

This lauded mollusc, he also states, is rather smaller than an oyster, which it does not resemble, being white, fat, and of unequalled flavour.

In case any of our readers shall fancy that fish diet may incapacitate them for laborious exertion, we add the following observations of Mr Yarrell:—" Two physicians of great experience, at the west end of London, find their advantage in taking fish only at dinner, almost daily. A surgeon, in extensive practice at the West End, tells me that whenever he foresees he shall have a hard day's work, and be obliged to go out in the evening, he arranges for an early dinner of fish, and finds that he gets through a hard day with much more ease to himself than if he had taken a heavy dinner of meat. A case of local inflammation, which resisted all the usual means, gave way entirely after persevering in a fish diet for nine weeks, and wholly abstaining from meat."

Upon the whole, therefore, we trust that our readers will agree that our articles on Pisciculture are of importance, because directing attention to the domain of the waters, from which the fishermen of the British Islands draw an amount of human food equal to little less than a million of money—a sum capable of indefinite enlargement, if those interested in fisheries will only combine to protect them from the ruinous effects of ignorance and shortsighted cupidity.

Oyster and mussel culture open up a new field of most profitable industry, which, we trust, will not long be left unoccupied. The interests of the public economy, as well as of large sections of our maritime communities, are so deeply implicated in our carrying out the suggestions of M. Coste, that we have been at pains to make them known. The expense to be incurred is comparatively trifling; and the details we have given in reference to the falling away in the supply of oysters, in particular, as well as of mussels in certain localities, render it too evident that we are in danger of witnessing a rapid diminution of both these molluscs—the one of which is as welcome to the epicure as the other is indispensable to the fisherman.

We have not been deterred from commending *ostreoculture* by the fierce abuse of the old author who maintains that "oysters are ungodly, uncharitable, and unprofitable meat—ungodly, because they are eaten without grace; uncharitable, because they have nothing but shells; and unprofitable, because they must swim in wine." The last assertion is in notorious opposition to that epicurean oracle, Dr Kitchener, who speaks thus:— "Those who wish to enjoy this delicious restorative in its utmost perfection must eat it the moment it is opened, with its own gravy, in the under-shell. The true lover of an oyster will have more regard for the feelings of his little favourite than to abandon it to the mercy of a bungling operator; he will always open it himself, and contrive to detach the fish from the shell so dexterously that the oyster is hardly conscious he has been ejected from his lodging." "Oysters," continues this famous gastronomer, "being of a mild, balsamic, and cooling nature, are peculiarly adapted as an article of food to those who are subjected to face-flushings, and other feverish symptoms, appearing in nervous and irritable, or consumptive constitutions." The number of red-faced, nervous, or irritable, not to mention the consumptive, is so great in these northern regions of ours, that there must be a constant demand for the myriads of oysters which our piscicultural papers will, we trust, be the means of producing. And if, as is too probable, some of our readers, yielding to the seductions of the 'Cook's Oracle,' overload their stomachs with savoury morsels, we have charity enough to give them this prescription to relieve their misery: "If oysters are felt lying too cold or heavy on the stomach, in consequence of too many having been eaten, a pint of new milk, taken immediately, will dissolve the oysters into a cream-jelly, and thus dissipate all symptoms of indisposition. Persons who are weak or consumptive should always take this after a meal of oysters."

Having thus, in the sayings of Dr Kitchener, furnished an antidote to the probable effects of his too seductive eloquence, we hope that our readers are prepared for the coming of the oyster-times foretold by M. Coste.

PEARLS AND PEARL-CULTURE.

WHAT are pearls, and where produced? Can they be multiplied by the art of pisciculture? These are queries of some importance, if it be true that, so recently as last century, the pearls sent to London from the rivers Tay and Isla between the years 1761 and 1764 were worth £10,000.* It is singular, observes Mr Tytler, "to find so precious an article as pearls amongst the subjects of Scottish trade; yet the fact rests on good authority. The Scottish pearls in the possession of Alexander I. were celebrated in distant countries for their extreme size and beauty; and, as early as the fourth century, there is evidence of a foreign demand for this species of luxury." The fame of our Scottish rivers as producing valuable pearls has of late been attracting some attention. Sums varying from a few shillings up to £40 have recently been given for single pearls; and it is stated that her Majesty and other persons of distinction willingly purchase pearls obtained from the Tay and the Don. Mr Unger, a jeweller in Edinburgh, is exerting himself in stimulating the people on the pearl-producing rivers of Scotland to cultivate this kind of industry. We have thus been led to think about pearls, and to investigate into the causes of their production: the result is, that we think pearl-culture deserving of attention. If oysters and mussels be

* Forbes and Hanley, 'British Molluscs,' ii. 149.

reared as articles of food, it is worth while to consider whether the bivalved molluscs producing pearls may also become the subjects of improved culture. As there is evidence that they may, we think it may interest some of our readers to make them acquainted with it.

But we have first to consider the vexed question, What is a pearl? According to an old popular fancy, pearls are dew-drops, which are transmogrified thus, according to Boethius: "Early in the morning, the mussels, in the gentle, cleare, and calme aire, lift up their upper shells and mouthes above the water, and these receive of the fine and pleasant breath or dew of heaven; and afterwards, according to the measure and quantitie of this vitall force received, they first conceive, then swell, and finallie produce the pearle." Poets having taken up this fancy, it survived until the researches of Mr Gray and Sir Everard Home demonstrated that pearls are merely the internal pearly coat of the shell, which, from some extraneous cause, has assumed a spherical form. It is supposed that the shell of the mollusc having been penetrated by one of its enemies, the creature repels the invader by the secretion of nacre; or, where sand or other small bodies have been accidentally introduced, they are covered with nacre as a protection against the irritation caused by their presence. A pearl having been once formed, the animal continues depositing concentric layers of fresh nacre. The pearls are usually of the colour of the part of the shell to which they are attached. "I have observed them," says Mr Gray, "white, rose-coloured, purple, and black, and they are sometimes said to be of a green colour. Their lustre, which is derived from the reflection of the light from their peculiar surface, produced by the curious disposition of their fibres, and from their semi-transparency and form, greatly depends on the uniformity of their texture and colour of the concentric coats of which they are formed."*

* 'Annals of Philosophy,' 1824.

Sir Everard Home prides himself as the discoverer of the true nature of the pearls. "I shall now explain what a pearl really is; and if in the course of my explanation I shall prove that this, the richest jewel in a monarch's crown—which cannot be imitated by any art of man, either in the beauty of its form or the brilliancy and lustre produced by a central illuminated cell—is the abortive egg of an oyster enveloped in its own nacre, of which it annually receives a layer of increase during the life of the animal, who will not be struck with wonder! In my investigation of the mode of breeding of the oyster and mussel, when the ova were examined in the microscope, we commonly found round hard bodies, too small to be noticed by the naked eye, having exactly the appearance of seed-pearls, as they are called, in the ovarium, or connected with the surface of the shell in contact with the membrane covering it; which led me to consider this to be the situation in which pearls are originally formed, more especially as here they were not only very small, but uniformly of the same size, and when found more and more distant from this spot they had increased in size. These facts led me to conclude that the ova which prove abortive do not die and drop off at the same time that those which have been impregnated pass into the oviduct, but remain in their capsulæ, which, being still supplied with blood-vessels, go on increasing for another year; their surface then receives a nacral covering with all the other surfaces of the shells, and they lose their attachment or become imbedded in the shell; this is in some measure proved by pearls being met with perfectly spherical—others in which the pedicles are included in the nacral coat—others, again, more or less buried in the nacral coat of the shell.

"As pearls have their origin from so small a nucleus, it is not surprising that there are so few of a large size, since it is probable that oysters of great size, which will depend upon their age, live in deep water beyond the reach of man." *

This conclusion of the comparative anatomist is not acquiesced in by writers on natural history.

"This," observes Mr Johnston, "is far from being true. I will not deny that the fact may be, in not a few instances, as stated by Sir E. Home—for the ovum may accidentally fall into a situation where it shall become a source of irritation, like any other extraneous substance; but that it is often the case is contradicted by numerous observations, and by the true theory of the formation of pearls. Professor Baer of Königsberg, aware of Home's theory, undertook an investigation of it in the fresh-water mussels of Germany; and the result was, that he never met with pearls in the ovaries, liver, kidney, or any of the internal organs. The pearls were always situated either in or under the skin of the back, where it is close to the shell."

This is certain—the animals producing pearls possess the power of covering with concentric layers of nacre portions of their shells needing to be strengthened, or objects introduced accidentally or by design. The Chinese have long practised the art of stimulating the secretion of nacre by the introduction of mother-of-pearl roughly filed into a plano-convex form, like the top of a mother-of-pearl button. Mr Gray, observing in the British Museum some very fine pearls thus produced, tried the experiment of introducing similar pieces of mother-of-pearl into the shell of the *Anodonta Cygneus* and *Unio Pictorum*. "If," he observes, "this plan succeeds—which I have scarcely any doubt it will—we shall be able to produce any quantity of as fine pearls as can be procured from abroad."

It is to be regretted that the result of this experiment is not known. As it deserves to be repeated, we give Mr Gray's account of the *modus operandi:*—"I found

* 'Comparative Anatomy,' vol. v. p. 313.

the introduction of the basis of the pearl attended with very little difficulty, and, I should think, very little absolute pain to the animal; for it is only necessary that the valves of the shell should be forced open to a moderate breadth, and so kept for a few seconds by means of a stop, and that then the basis should be introduced between the mantle and the shell by slightly turning down the former part and pushing the pieces to some little distance by means of a stick, when the stop may be withdrawn and the animal will push the basis into a convenient place by means of its foot; and of the thirty or forty bases which I thus introduced, only one or two were pushed out again, and these I do not think had been introduced sufficiently far."

In order that the experimenters in this mode of pearl-producing may not be disappointed with the result, we add these observations of Mr Gray:—" In cutting these pearls from the shell, it is necessary that the shell should be cut through, so that the mother-of-pearl button may be kept in its place; for if the back were removed, as it would be if the shell were not cut through, the basis would fall out, and then the pearl would be very brittle. The only objection that can be adduced against these pearls is, that their semi-orbicular and unequally-coloured sides preclude them from being strung or used any other way than set; but this fault will always be the case with all artificially-produced pearls, as the mantle can only cover one side of them; and the only pearls that will answer the purpose of stringing are those found imbedded in the cells in the mantle of the animal."

In the 'Encyclopædia Britannica,' vol. vi. p. 477, we find this notice of the way of producing pearls artificially:—" The shell is opened with great care, to avoid injuring the animal, and a small portion of the external surface of the shell is scraped off. In its place is inserted a spherical piece of mother-of-pearl about the size of a small grain of shot. This serves as a nucleus on which is deposited the pearly fluid, and in time forms pearl."

Linnæus was persuaded that pearls could be produced at pleasure by simply puncturing the shell with a pointed wire; and pearls formed in this manner are preserved in the Hunterian Museum.

But enough as to the mode in which pearls may be formed. The process of their artificial formation is so simple that it deserves to be the subject of further experiment.

What are the creatures by which pearls are produced? They are chiefly bivalved molluscs—the edible mussel, the fresh-water mussel, and what is improperly termed the pearl-oyster. Pearls are also found in the oyster of the British coasts; and we have three found in an oyster presented at table in Edinburgh. The greatest pearl-fishery in the world is off Ceylon, which at present is of especial interest, owing to Government having resorted to the formation of artificial oyster-banks, suggested by the method recently introduced along the shores of France for repeopling the wellnigh ruined banks of the edible oyster. It is necessary to have recourse to artificial banks, it being found that a remunerative annual fishery cannot be obtained, owing to the banks being exposed to the violence of ocean currents, which, by washing sand into the interstices of the rocks, often destroy the young oysters over a considerable area. The oysters, moreover, are exposed to the pernicious influence of the *soorum*, a species of *modiola* like a mussel with a swollen face. The Tinnevelly pearl-banks, which in 1861 yielded 15,874,500 shells, realising a profit of more than £20,000, were in 1863 found to be in so unpromising a condition that no fishery was attempted. It is to remedy the uncertainty of this valuable fishery that Dr Kelaart, after investigating the nature of the pearl-oyster (*Meleagrina margaritifera*), declares it as his opinion that he "sees no reason why pearl-oysters should not live and breed in artificial beds like the edible oysters, and yield a large revenue." He has ascertained, by his experiments in Ceylon, that the pearl-oysters are more tenacious of life than any other bivalve with which he is acquainted, and that they can live in brackish water and in places so shallow that they must be exposed for two or three hours daily to the sun and air.

Captain Phipps, superintendent of the Tinnevelly pearl-banks, convinced that artificial nurseries for the young oysters are the only means to insure a remunerative fishery, has succeeded in getting these established on a bank in the harbour of Tuticorin.

In compliance with custom we have spoken of the pearl-oyster, but the animal is really a mussel, having, like it, a *byssus* or cable by which it secures itself to the rocks. Dr Kelaart's researches in Ceylon have proved that it possesses the power of casting off its byssus at pleasure. We trust that this only recently-ascertained fact will attract the attention of the Acclimatisation Societies in London and Paris, and induce them to remove the pearl-oyster from its native beds and locate it in the seas of Europe.

At all events, now that the former of these societies has accepted a gift of fresh-water mussels from the Tay, with the design of introducing them into an English river, we hope that something will be done to encourage the transportation of the mussel (*Unio margaritifera*) to the many rivers and lakes suitable to their production. The mussels of the Tay and the Don in Scotland, of the Conway and the Irt in England, and of several rivers in the North of Ireland, have long been known to yield pearls, often of great value. But the natural history of the animal has been little studied; its ability to bear transportation to a distance, the means of rendering it prolific, and of developing its pearl-producing powers by placing it in favourable localities, or by such artificial processes as we have indicated—all these are points about which little is known, but as to which we hope we shall soon know more. Should any of our readers be induced to search for pearls, we give them this hint from an old writer,—"The shells that have the best pearls are wrinkled, twisted, or bunched, and not smooth or equal, as those that have none." Should they fall in with any of good size and shape but deficient in colour, let them remember this observation of Sir Alexander Johnston, for many years President of His Majesty's Council in Ceylon:—"The pearl is composed of strata or scales which are easily removed by a skilful hand, without injury to that below, which retains all its brilliancy. Pearls of a large size and perfect form, but discoloured, have been bought at a low price, but became exceedingly valuable by removing one or two of the upper coats."

HORSES—ANCIENT AND MODERN.*

WE have made our readers acquainted with M. Isidore Geoffrey Saint-Hilaire's 'Letters on Alimentary Substances, and especially on the Flesh of the Horse,' and with our conversion to the faith of the *Hippophagi*. Our longing after the gastronomic treat of a "horse-dinner," naturally made us curious to ascertain, statistically, how much of this irrationally neglected species of food might be made to find its way into the too often empty flesh-pots of the people of this country. After due research we ascertained that in Scotland there were nearly 180,000 horses; and M'Queen's statistics informed us that in the United Kingdom there are 2,250,000 horses valued at £57,000,000. "The wish" being "father to the thought," we had, of course, no hesitation in arriving at the conclusion that a large proportion of these should be eaten by human beings.

The number of horses within these realms is a very important matter indeed, in these times when the value of horses is so greatly increased both at home and abroad. Fortunately for our present purpose, an article in the 'Mark Lane Express' furnishes information of a later date than that of our former article. With this we shall make our readers acquainted, in

* 'The Horse and his Rider.' By Sir Francis B. Head, Bart. London: John Murray. 1860.

order that they may have a distinct idea of the value and importance of the horse, dead or alive, as an article of food, or as the chief animal ally of man in peace or in war.

According to the census of 1857, there were in Scotland 185,409 horses, classed as follows:—for agricultural purposes, above three years old, 126,471; under three years old, 34,947; all other horses, 23,991. The number in Ireland in 1857 was 600,693; of which for agricultural purposes there were 16,606; for traffic and manufactures, 2466; for amusement and recreation, 2469; 1779 were yearlings, and 3965 were under one year. The value per head, estimated by the Census Commissioners of 1841 at £8 per head, gives an aggregate for Ireland of no less than £4,806,044.

We have no reason for doubting the accuracy of these statements; and yet, even after making allowance for the inferiority of Ireland to Scotland in respect to tillage, it seems incredible that in Scotland there should be 126,471 horses employed in agriculture, and in Ireland only 16,606. If our readers share in our surprise, we pray them to charge the incredibility not on us, but on the 'Mark Lane Express.'

For England and Wales, there being no specific return, it is estimated that there are at least 1,050,930 horses, and 258,079 colts. Independent of the value of the above two million and a quarter horses within the United Kingdom, the gross produce to the revenue for Great Britain, from the taxes on horses and horse-dealers, was, in 1859, £362,193. We cannot therefore question that an animal which produces to the public revenue nearly a thousand pounds per day must be an especial favourite with the Chancellor of the Exchequer, who doubtless groans in spirit that he cannot saddle with taxes more than a million or so of horses, whereas, if in Russia, he might swell out his budget by financial operations on some seventeen millions of these noble animals. These are chiefly to be found in the pro-

T

vinces of Oran and Perm, the inhabitants of which have a special aptitude for horse-breeding, and in the country of the Don Cossacks, where horsemanship is an indispensable part of the daily avocations of the people.

Our import and export trade in horses is important. In the six years preceding 1859 we imported 24,558, and exported 13,218; the average value of those imported being only about £20, while these exported were worth about £90 a-piece.

The Indian Government has long been paying great attention to the rearing of horses, as well as importing them from the Cape colony and Australia. Those purchased in Egypt, at about £25 each, and shipped to Bombay, it is calculated, cost the Government nearly £100 each, irrespective of casualties. The demand for horses, instead of being diminished by the introduction of railways, has been notably increased; so little truth has there been in the prophecy which appeared in the 'Quarterly Review' some twenty-five years ago, that the food of the horses supplanted by railways would suffice for the nourishment of several millions of people.

However interesting the preceding details may be to those engaged in the rearing of horses, and however proper it might be to dilate upon their importance to the interests of agriculture, we turn from them, because we do not intend to write a practical paper on horse-flesh—or, as our French neighbours scientifically phrase it, "the equine species." We are merely about to tell our readers a few things about horses in that gossiping style in which Sir Francis Head is such a proficient. There is no diminution in the vivacity, the wit, the varied knowledge of men and things first exhibited in 'Rapid Journeys across the Pampas and over the Andes,' and subsequently in 'Bubbles from the Brunnens of Nassau,' 'Stokers and Pokers,' 'The Defenceless State of Great Britain,' 'A Faggot of French Sticks,' &c. In 'The Horse and his Rider,' the "old man"—as he is fond of styling himself—is as witty and wise as ever, and really has a great deal to say well worth hearing. The human biped gets many a hard hit while apparently only hearing about the quadruped, in gratitude to which Sir Francis submits to the consideration of the public what he terms "imperfect observations applicable to all living creatures." Sir Francis, instead of quoting the common saying, "A merciful man is merciful to his beast," gives the words of Solomon, "*A righteous man regardeth his beast.*" As Solomon had no less than "40,000 stalls of horses for his chariots, and 12,000 horsemen," it is pleasant to know that in this instance, at least, the royal preacher observed his own precept, and that "his officers provided victual for King Solomon, *barley also and straw for the horses and the dromedaries.*" Knowing how frequent is the use of *barley* as food for the horse all over the East, we have often wondered at its little employment in this country for the like purpose.

The mention of Solomon "in this connection," as some folk oddly express themselves, has induced us to look into his horse-dealing business, and the result is that we ascertain that it was on a great scale, and so profitable as to excite the envy of many a dealer in our modern horse-markets. This royal *couper** "had horses brought out of Egypt, and linen yarn : the king's merchants received the linen yarn at a fixed price. And they fetched up, and brought forth out of Egypt, a chariot for six hundred shekels of silver, and a horse for a hundred and fifty; and so brought they out *horses* for all the kings of the Hittites, and for the kings of Syria, by their means" (2d Chronicles, i. 17). It is not easy to understand how a people so "learned" as the Egyptians should allow such a formidable neighbour as Solomon to acquire a large force of cavalry, and enrich himself by a monopoly in the trade of their horses. The German commentator Michaelis is of opinion that "the fixing of the price of the horses has the look of a

* *Anglice*, horse-dealer.

monopoly, and indicates, besides, that horsemanship was in its infancy; for whenever people have sufficient knowledge of horses, with all their combinations of faults and excellences, and learn to judge of them as *amateurs*, one individual of the same breed may be worth ten times as much as another, particularly in a king's stables." We think the learned German is right. Fancy-prices must have been unknown in Egypt when Solomon's factors had the choice of its horses at a price varying from £17, 2s. to £18, 15s., according to the lower or higher valuation of the shekel (2s. 3d. to 2s. 6d.) Having already stated that the present Government of India imports horses from Egypt at the rate of about £25 a-head, it is curious to observe that the average price of a horse in the East is apparently as unchanging as many of its other usages. Solomon (B.C. 1015) paid about £18, 15s.—the British Government last year was paying £20. As a piece of practical information, we beg our horse-buying readers, and especially those in search of hunters, to listen to Sir Francis Head : " A respectable first-rate horse-dealer succeeds in his profession, not so much by his superior knowledge of the animals he *buys*, as by the quantity and quality of the eloquence he exerts in *selling* them. Every hunter, therefore, that is purchased from a great man of this description is necessarily composed of—first, his intrinsic value; and, second, of the anecdotes, smiles, compliments, and praises, which—although, when duly mixed up with an evident carelessness about selling him, they captivate the listener to purchase him—like a bottle of uncorked ardent spirits, evaporate, or, like a swarm of bees, fly away, almost as soon as the transaction is concluded, leaving behind them nothing but the animal's intrinsic value."

Though the laudations of a modern horse-dealer are always to be taken at a large discount, those of an Arab for his favourite steed are genuine. The passionate admiration of the Arabians for the beauty, strength, and intelligence of the horse is thus truly indicated in the 'Romance of Antar.' " Shedad's mare was called Jirwet, whose like was unknown. Kings negotiated with him for her, but he would not part with her, and would accept no offer or bribe for her, and thus he used to talk of her in his verses: 'Seek not to purchase my horse, for Jirwet is not to be bought or borrowed. I am a strong castle on her back, and in her bound are glory and greatness. I would not part with her were strings of camels to come to me with their drivers following them. She flies with the wind without wings, and tears up the waste and the desert. I will keep her for the day of calamities, and she will rescue me when the battle-dust rises.'"

D'Arvieux relates a story of an Arab of Tunis who would not deliver up a mare bought for the stud of the King of France. "When he had put the money in his bag, he looked wistfully on his mare, and began to weep. 'Shall it be possible,' said he, 'that after having bred thee up in my house with so much care, and after having so much service from thee, I should be delivering thee up in slavery to the Franks for thy reward? No! I will never do it, my darling.' And with that he threw down the money, embraced and kissed his mare, and took her home with him again." In the Rev. V. Monro's 'Summer Ramble in Syria' we meet with a more recent instance of this friendship between the horse and his rider:—"A great ruffian was mounted on a white mare of great beauty. Having asked her price, I offered the sum. The Arab said he loved his mare better than his own life, that money was of no use to him, but that, mounted on her, he felt rich as a pasha. Shoes and stockings he had none, and the net value of his accoutrements and dress might be calculated at something less than seventeenpence sterling."

Very great dubiety exists as to the native region of the horse. Geologists inform us that his remains are found in nearly every part of the world. "His teeth lie in the polar ice along with the bones of the Siberian mammoth; in the Himalaya Mountains with lost and but recently obtained genera; in the caverns of Ireland; and, in one instance, from Barbary, completely fossilised. His bones, accompanied by those of the elephant, rhinoceros, tiger, and hyena, rest by thousands in the caves of Constadt, in Sevion at Argenteuil, with those of the mastodon in Val d'Arno, and on the borders of the Rhine with colossal urus."*

It is surely in consequence of heavenly providence caring for man, that while these creatures, found under the same conditions, have ceased to exist, or have removed to higher temperatures, the horse remains in the same regions, without, it would appear, any protracted interruption, "fragments of his skeleton continuing to be traced upwards, in successive formations, to the present surface of the earth—the land we live in." His appearance, however, on the continents of the New World, whether of the Atlantic or Pacific Ocean, is of comparatively recent date.

This wide diffusion of the horse in all countries renders it the more difficult to determine his original *habitat*. Naturalists incline to think that it was Arabia; but until the time of Joshua (B.C. 1450) Dr Kitto maintains that we find in Scripture no mention of the horse unless in connection with Egypt. He forgets, however, that Jacob, a native of Palestine, but dying in Egypt (B.C. 1689), compares Dan to "an adder in the path, that biteth *the horse heels*, so that his *rider* shall fall backward," Gen. xlix. 17. Moreover, the sublimest description of the horse is in the Book of Job, which the best scholars refer to the epoch of the patriarchs; thus making it the oldest book in the world. And as Job dwelt in the land of Uz, "the daughter of Edom," we identify his country with that of "the land of Edom," in Arabia Petræa. "Hast thou given the horse strength? Hast thou clothed

* Naturalists' Library, vol. xii.

his neck with thunder? Canst thou make him afraid as a grasshopper? The glory of his nostrils is terrible. He paweth in the valley, and rejoiceth in his strength; he goeth on to meet the armed men. He mocketh at fear, and is not affrighted, neither turneth he back from the sword. The quiver rattleth against him, the glittering spear and the shield. He swalloweth the ground with fierceness and rage, neither believeth he that it is the sound of the trumpet. He saith among the trumpets, Ha, ha! and he smelleth the battle afar off, the thunder of the captains, and the shouting." Michaelis is of opinion that none but a military man, who has observed the war-horse in battle, can fully appreciate the force of this description. "I have," he observes, "rode more perhaps than many of those who have be-become authors, or illustrators of the Bible; but one part of the description—namely, the behaviour of the horse in the attack of a hostile army—I only understand rightly from what old officers have related to me; and as to the proper rendering of the two lines, 'Hast thou clothed his neck with ire?' ('with thunder,' in our version), and 'the grandeur of his neighing is terror' ('the glory of his nostrils is terrible,' in our version), had escaped me; indeed, the latter I had not understood, until a person who had an opportunity of seeing several stallions together instructed me, and then I recollected that in my eighteenth year I had seen their bristled-up necks, and heard their fierce cries when rushing to attack each other." We daresay our readers will be glad that we add to this the testimony of Sir Francis Head as to the intrepidity of the horse: "As soon as his courage is excited, no fall, bruise, blow, or wound, that does not paralyse the mechanism of his limbs, will stop him; indeed, with his upper and lower jaw shot away, and with the skin dangling in ribbons, we have seen him cantering, apparently careless and unconscious of his state, alongside of the artillery-gun from which he had just been cut adrift."

But in order that the horse may not suffer under the imputation of being merely an excitable brute, roused to frenzy by "the pomp and circumstance" of war, we must allow Mr Youatt to testify how lasting is the sympathy of this generous creature for those of his own kind: "In some the friendship is so intense that they will neither feed nor live when separated from each other. Two Hanoverian horses had long served together during the Peninsular war: they had drawn the same gun, and had been inseparable companions in many battles. One of them was at last killed; and after the engagement was over, the survivor was picketed as usual, and his food brought to him. He refused to eat, and was constantly looking about in search of his companion, sometimes neighing as if to call him. He was surrounded by other horses, but he did not notice them; and he shortly afterwards died, not having tasted food from the time when his former associate was killed."

We have rode off at a tangent. Coming back to our starting-point, as to the native country of the horse, there really appears to be no sufficient ground for believing it to be Arabia. We have a complete inventory of the live-stock of Job and of several of the patriarchs, but the horse is mentioned in none of them; and when the Jews were brought into contact with the nomades of Arabia, they found "their camels were past numbering," and that even their kings rode on camels (Judges viii. 21). Ancient history makes no allusion to Arabia as distinguished for its horses; and Strabo, who wrote so late as the time of Christ, expressly declares that it was destitute of these animals. The Arabs trace the genealogies of their best horses to Solomon's stud; but this is as manifestly fabulous as their tradition of their own descent from the Queen of Sheba's *liaison* with the wise king. It is certain that by the time of Mahomet the Arabs had paid attention to the breeding of horses, but when they began to do this is unknown. As to the country from which they obtained their breed, we can hardly doubt that it was Egypt. We have already seen that the art of horsemanship was known so early as the time of Jacob. Profane historians describe it as an Egyptian invention, assigning it to Osiris himself. The Egyptian monarchs appear to have prided themselves in having vast numbers of horses. Diodorus mentions that the kings before Sesostris had a hundred stables, each for two hundred horses, on the banks of the Nile between Thebes and Memphis. That they possessed an esteemed breed of horses is evident from Egyptian paintings; which, however, all corroborate those ancient writers who affirm that the horse was not used in agriculture, but appropriated to the pomp of luxury and war.*

Whatever be the native country of the horse, let us be thankful that we have him in as great variety as any nation, and that we are a race of bold riders, well knowing how to draw forth his noble qualities, and ride him with an ease and firmness of seat in gratifying contrast to the unstable position of a French equestrian.

The first requisite in horsemanship is intelligent courage, the second being a *just* seat. The attitude assumed by civilian riders throughout the United Kingdom is what is called "the hunting-seat," in which, instead of "the fork," the knees form the pivot, or rather hinge, the legs beneath them the grasp, while the thighs, like the pastern of a horse, enable the body above to rise and fall, as lightly as a carriage on its springs. Though facility of turning in the saddle be not so easy in this attitude as when the rider revolves upon his "fork," Sir Francis assigns various good reasons for preferring it. "One of the most usual devices by which a horse endeavours to dislodge his rider is by giving to his back, by a sudden kick, a jerk upwards, which, of course, forces in the same direction towards the sky that nameless portion of humanity which was partly resting on it, and which in the cavalry cannot possibly

* Rayner—'Economie Publique et Rurale des Egyptiens.'

get very far away from it. But in the hunting-seat, the instant a rider expects such a kick, by merely rising in his stirrups he at once raises from the saddle, the point his enemy intends to attack, and accordingly the blow aimed at it fails to reach it."

The worst of positions is the bent attitude of the last paroxysm or exertion which helped the rider into the saddle. The first bad trip even projects him over his horse's head in a parabolic curve, ending in a concussion of his brain or the dislocation of his neck, the horse being uninjured. But when a man sits in his saddle justly balanced, a sudden jerk forwards throws his shoulders backwards. "In the event of a fall, the horse is the only sufferer. He cuts his forehead, hurts his nose, breaks his knees, bruises his chest; while his head, neck, fore-legs, and the fore-part of his body, forced into each other like the joints of a telescope, form a buffer, preventing the concussion the horse has received from injuring, in the smallest degree, the rider. Seated in his saddle in the attitude we have described, that admirable rider Jack Shirley, whipper-in to the Tedworth hunt, with a large open clasp-knife in his mouth, was one day observed fixing a piece of whipcord to his lash, while following his hounds at a slapping pace down hill, his reins lying nearly loose on old Gadsby's neck."

In confirmation of the theory that, when a man sits properly in his saddle, it is the horse, and not the rider, who suffers by a tumble, we are referred to several remarkable escapes. A Northamptonshire rider, taking a fence, jumped over it into a stone quarry. If he had been in the bent attitude already alluded to, he must have pitched on, and broken his skull, whereas his seat being "just," only his ankles suffered. In like manner, when, to escape the murderous stratagem of Mehemet Ali, Amyn Bey spurred his charger over the low wall of the citadel at Cairo, instead of being crushed by falling on the rock fifty feet below, he was able to crawl away with only a broken ankle, while the horse was dashed to pieces.

A still more extraordinary escape was that of General Yorke, as thus related by himself: "In June 1848, at the island of Dominica, in the West Indies, I fell over a precipice of 237 feet perpendicular height, upon the rocks by the seaside. Every bone of my horse was broken, and I conceive my escape from instant death the most miraculous that ever occurred. My recovery from the shock I sustained was almost as miraculous as my escape with life. I sent out an artist to take a drawing on the spot, and also had the place surveyed by an engineer. I have often thought of putting down all the circumstances of that extraordinary accident, but the dread of being taken for a Baron Munchausen has restrained me. I do not expect that any one will believe me, although there are many living witnesses. Nor do I expect any sympathy, for, as soon as I could hold a pen, I detailed the catastrophe to my mother, to account for my long silence. I received in reply, in due course, a long letter detailing family news, without any allusion to my unfortunate case, except in a postscript, in which she merely said, '*Oh, William, I wish you would give up riding after dinner.*'

"*P.S.*—The accident occurred *before* dinner. During the fall I stuck to my horse."

This incident illustrates at once the sagacity and the intrepidity of the horse. He was urged by his rider to take the fatal leap; but not till after being several times forced, almost at full speed, to approach an obstacle which inspired him with dread, did the poor animal do violence to his instinct, and bound over the frightful precipice, at the foot of which he was found literally smashed to atoms.

In the great plains of South America the rider sits almost perpendicularly, with the great toe of each foot resting very lightly, and often merely touching, its small triangular stirrup, his legs grasping the horse's sides slightly or tightly, as prosperous or adverse circumstances may require. We are assured that this attitude is not only highly picturesque, but particularly easy to the rider, who, while partaking of the undulating motion of the horse, can rest his wearied body by slight imperceptible changes of position on the pivot or fork on which it bends. The British cavalry ride very nearly in this position, which, as Sir Francis Head explains, enables them with great facility to cut or give point in front, right or left; and "if they were not embarrassed by their clothing as well as by their accoutrements, and if, as in South America, they were to use no pace but the gallop, each would soon become apparently part and parcel of his horse. But our gallant men continue not only hampered and imperilled by a horse-cloak, holsters, and carbine in *front* of their thighs, and imprisoned, especially round their necks, within tight clothing, but their travelling pace, the trot (a jolting movement unheard of in the plains of South America), gives to their body and limbs a rigidity painful to look at, and in long journeys wearisome to man and horse. Indeed, in the French cavalry, and occasionally in our own, the manner in which the soldier, in not a bad attitude, is seen hopping high into the air, on and off his saddle, as his horse, at apparently a different rate, trots beneath, forms as ridiculous a caricature of the *art of riding* as the pencil of our *Punch's* 'Leech' could possibly delineate."

We lately, with compassionate mirth, beheld a regiment of hussars thus equestrianising through the streets of Edinburgh; and certainly the droll look of the thing we have rarely seen exceeded, even in the case of a farmer returning from market in that peculiar style in which a man is said to be " on the outside of his horse," an external appendage connected with the animal in a manner ludicrously unstable. Like everything in nature the variety of seats is infinite; but Sir Francis is unable, with all his cleverness, to account for the phenomenon of the French, who excel us all in dancing, walking, and fencing, being the worst riders in Europe. "In all other countries, a man, grasping more or less firmly with his knees his saddle-flaps, allows his body freely to partake of the motion of the horse, until, with our best riders, the two, as they skim together over rough ground, appear to form one animal. In France, however, the rule is diametrically the reverse, for the moment the horse begins to canter, the rider's legs become like a pair of scissors astride an iron poker, and, while they appear useless, his back assumes the shape of a new moon. In fact, the French have no more seat on a horse than a parched pea has on a shovel; and as they trot along, popping up and down at one pace, while their fine English quadruped is boldly striding onwards at another, I have constantly expected to see even a dragoon trotting along with a despatch, hop, hop, hop, hop over the tail to his mother earth. In short, their uncomfortable appearance always reminds me of the toast proposed by an inhabitant of the state of Mississippi—'Gentlemen, I give ye a high-trotting horse, cobweb breeches, and a porcupine saddle, for the enemies of our glorious institutions!'" In his 'Faggot of French Sticks,' from which this is extracted, we learn that Sir Francis had the honour of accompanying Louis Napoleon, while President of the Republic, to a grand review. "As soon as we passed the Bridge of Jena, the President, who proverbially, in France, is '*parfaitement bon cavalier*,' started off in a gallop; and accordingly, between the troops drawn up in line, and whose bands successively struck up as we reached them, we had a scurry across the Champ de Mars which was really quite delightful; indeed, my horse seemed so pleased with it, that, had it not been for my curb-rein, I believe, very much against my will, he would have 'come in first.'"

Whether dining and riding with an emperor might mollify our suspicions of Louis Napoleon we shall not stop to consider. It so happens, however, that we do not share in the admiration of Sir F. Head for the "benevolent countenance" and "good heart" of this inscrutable potentate—to such a degree, at least, as to deem it safe to intrust our liberties to the forbearance of the lord of so many legions of the most impulsive and fearless soldiers in Europe; and therefore, instead of following Sir Francis to the hunting-field, and bringing before our readers his graphic sketch of a "meet," and his many wise and humorous remarks on "the horse and his rider," when engaged in a sport which calls forth so remarkably the characteristics of both, we turn to the chapter on "Military Horse-Power." The success of an army in the field, most people fancy, mainly depends upon the attention paid to the victualling department. Of course, with little food in his stomach, the soldier has little stomach for hard fighting and long marches. And if unfit for long and rapid marches, an army is comparatively useless, for, according to Marshal Saxe, "its arms are of less value than its legs." The destructive power of the army has been amazingly increased, but at the expense of its activity; so that a European, like an East Indian army, often finds it easier to fight than to march. This fatal defect is ascribed to the non-development of the physical power of the horse. Guns, ammunition, treasure, &c., which European cavalry have bravely won, their horses have been supposed unable to carry away. And in great emergencies, when infantry, oxen, and mules were all taxed to the utmost in accumulating means for attacking a fortress, the cavalry have not borne their part in the burden and heat of the day, because, "although it is easy to extract from men manual labour, it is concluded that it is impossible to extract from horses horsepower." This is illustrated by the practice of the horse-artillery during the Peninsular War. To each gun there were attached twelve horses trained to draught. Of these only eight possessed the means of drawing.

If their power proved insufficient to overcome an obstacle which they chanced to encounter, the gun was either deserted, or dragged along by the infantry, four draught-horses all the time standing by doing nothing. Nobody can dissent from the opinion that, in common life, it is eminently silly to suffer from not knowing how to use what we actually possess; and therefore it must be granted that the long delay in calling forth the auxiliary strength of the horse is a blot upon the military character of those engaged in war. It is undeniable that the science of war demands the putting forth of the maximum strength and activity alike of cavalry and infantry; and bearing in mind that recent improvements in gunnery make the fate of battles depend upon the celerity with which artillery can be brought to play upon particular points, it evidently is a matter of prime importance that the horse-power attached to an army should be carefully economised and skilfully directed.

Influenced by such considerations, the Duke of Cambridge lately made this addition to the Army Regulations:—" In order that the cavalry may, upon emergencies, be available for the purposes of draught—such as assisting artillery, &c., through deep roads, and in surmounting other impediments and obstacles which the carriages of the army have frequently to encounter in the course of active service—ten men per troop are to be equipped with the tackle of the *lasso*."

So, then, two traces, as means of traction, are henceforth to be at a discount; and horses, blinded by these absurdities, blinkers, which hinder them from knowing what they are called on to do, and fill their heads with chimerical terrors, are phenomena about to disappear. Henceforth our excellent ally, the horse, is, with his eyes open, to be called on to bear an equitable share in the toils of her Majesty's armies; and this by the simple expedients of a surcingle and a single trace—by which means have been transported, time out of mind, all the wheeled carriages conveying men and merchandise across the vast plains intervening between the oceans of the Atlantic and the Pacific.

In his 'History of the War of Independence,' General Miller attests their efficiency as means of military transport:—" Our corps consisted of ten six-pounders and one howitzer. Each gun was drawn by four horses, and each horse ridden by a gunner, there being no corps of drivers in the service. A non-commissioned officer and seven drivers were, besides the four already mentioned, attached to each piece of artillery. Buckles, collars, cruppers, and breastplates, were not in use; the horses simply drew from the saddle, and with this equipment our guns have travelled nearly one hundred miles in a day.

This vision of horses, *minus* blinders and buckles, collars and cruppers, may grieve the soul of a saddler. When carried out in the British army it will assuredly rejoice the horse and his rider, and materially aid in the achievement of victory.

We have long been convinced that blinders are dangerous absurdities. Should they become detached from the poor beast which bears them, he is suddenly aware of the, to him, alarming appearance of a man on the top of a hill, attached in some inexplicable fashion to his tail; or of a strange hairy animal similarly attached, and vomiting smoke and fire. The hill is a load of hay, and the hairy horror is a hirsute hunter, smoking his cigar on his way home from the hounds. But the horse does not know that; and so, when the blinders are suddenly removed, and he sees objects so unusual, he is seized with a fit of terror which prompts him to flee from them at his utmost speed, which is not diminished by his perceiving that somehow they follow him with equal velocity.

It cannot be questioned that if the animal had been treated on the rational system of Mr Rarey—that is, if he had been allowed to see, smell, and touch the vehicles to which he was harnessed—his composure would not have been disturbed.

We have tried the experiment on a gig-mare accustomed to blinders. On first seeing the gig she was alarmed, but, being encouraged by us, as well as by a strong man alongside of her in case of a bolt, she very soon proceeded with her usual propriety.

As to the lasso and the surcingle, we have no experience of them; but as it is demonstrated, by the recent experience of the Royal Engineer Train, that with this simple equipment any number of horses, *whether accustomed to draught or not*, can be harnessed to any kind of carriage, not only in front, but in rear, to hold it back, or even sideways to prevent its oversetting; as this is the common mode of draught in South America for all purposes; as the traces and surcingle are there made of nothing but bullock-skins (costing less than English girths and surcingles),—it is evident that this mode of traction would be most serviceable on the far better roads and bridges of this country. And why should not the farmer have the benefit of this simple and effective mode of applying the horse-power on his farm? Why should *two* horses be overburdened doing what could be better and more expeditiously done by four or six exerting their strength in the most advantageous manner by means of single traces? We must leave our agricultural friends to determine the occasions of using the single trace; that they could often employ it profitably we are convinced. Sir Francis Head truly observes, that the many curious, and indeed scientific, applications and combinations of power of which this simple harness is capable, form a beautiful example of what even uncivilised man can contrive when his attention has been long and steadily directed to a solitary object. And surely the ingenuity and practical experience of one nation are worthy the patient attention of another. But though any horse will draw by the single trace, the rider must be experienced and attentive, otherwise the trace, in turning, will get under the horse's tail; and if at starting it be not properly held in the hand, it may be broken by a sudden jerk.

Sir Francis Head is kind enough to furnish us stay-at-home folks with other valuable foreign notions as to *anchoring* and *hobbling horses*. In South Africa, farmers and sportsmen " anchor " their horses by a lump of lead, from three to five pounds in weight, carried in a small pocket buckled to the outside of their near or left holster. To this is attached a piece of cord ten feet long which, passing and running freely through both rings of the curb-bit, and hanging from them like a loose rein, is fastened to a D or ring in the off-side of the saddle. If a horse thus anchored attempts to move on, his nose is brought down to his breast by the cord, which, tightening equally on both sides, acts exactly like a bridle in the hand of a rider; and as the pressure of the curb-chain ceases when he stops, he soon discovers that the best thing he can do is to stand still and graze. This invention has proved to be admirably adapted for farmers, for hunting and shooting, or for Staff and Engineer officers while reconnoitring or surveying.

Not less worthy of consideration is the process of *hobbling*, which, it appears, is alike serviceable in making war, or in the far more agreeable and harmless process of making love. In the countries of South America, every cavalry soldier carries a pair of hobbles, not in his pocket, but as an ornament dangling from the throat-lash of the bridle. " Whenever a young dandy calling on his *inamorata* is informed that she is 'in casa' (that is, at home), he dismounts, extracts from his waistcoat pocket a beautiful pair of silver hobbles (weighing only two ounces), which by two silver buttons he affixes to the fetlocks of his high-bred horse, who, switching with his long tail the innumerable flies that assail him, and looking at every animal that canters by him, stands stockstill, until within the house all the compliments of the season have been paid, and all the songs to the guitar exhausted."

At the suggestion of Sir F. Head, hobbling has been adopted by the mounted troop of the Royal Engineer Train; and any one visiting Aldershot is now enabled to see six or eight horses hobbled at intervals of about thirty feet, standing motionless, while the riders of the rest of the troop to which they belong, with drawn sabres flashing in the sun, are galloping through them backwards and forwards. As cavalry horses could be made to do the same, that noble branch of our army, as well as our yeomanry cavalry, would be able to act as *mounted infantry* by merely carrying hobbles weighing a couple of ounces.

Let not our readers deny themselves the pleasure and the profit of perusing 'The Horse and his Rider,' under the notion that an old officer like Sir Francis Head can tell them nothing that they do not already know. Are they acquainted with "hobbles and anchors?" with the best mode of swimming a horse? with the use and abuse of spurs? with how to bring a hunter home? with the fact that, though a groom ought to keep his horse clean, his primary study is to clean his stable? that it is almost impossible to keep straw under a horse perfectly pure, and that throughout the United States of America, and even in New York, horses are often made to lie on bare boards, on which they seem to sleep as soundly as, in a state of nature, they certainly do on the ground baked hard by the sun? Or have they ever heard of how much damage and suffering to horses are prevented by the unilateral system of shoeing, which Mr Turner, of Regent Street, London, terms "half-nailing," which consists in affixing the shoes by nails in the outside and round the toe only, leaving the inner side totally unsecured? Do they know the merits of "Fitzwilliam girths?" Are they, in short, as wise as Sir Francis Head? who, he tells us, sometimes from private inclination, and sometimes for the benefit of his health, sometimes for recreation, sometimes to risk his life, and more than once to save it, has, throughout a long and checkered career, had to do an amount of rough-riding a little larger than has fallen to the lot of many men. We have our doubts! and therefore exhort them one and all to have modesty enough to own that they may pick up not a few valuable hints from so old and so shrewd a campaigner.

At all events, we cannot conceive any one of ordinary feeling not heartily sympathising with his vigorous denunciation of the horrible cruelties of "vivisection," as practised at Alfort and Lyons, the chief veterinary colleges in France, the pupils at which, twice a-week for seven hours a-day, are instructed in surgery by the cutting-up of living horses, which, until they actually expire, are subjected to unmentionable cruelties. Are we blameless in our treatment of horses when in the hands of the farrier or the veterinary surgeon? Far from it, when, *without the application of chloroform*, we permit such operations as cutting off and cauterising their tails, burning their sinews with red-hot irons, dividing and cutting out a portion of a nerve, and other excruciating operations on young horses, under which they are often heard to squeal from pain. "You are a man of *pleasure*," says Sir Francis—"save your horse from unnecessary pain. You are a man of business—inscribe on that ledger, in which every one of the acts of your life is recorded, on one side how much *he* will gain, and on the other how very little *you* will lose by the evaporation of a fluid that will not cost you the price of the shoes of the poor animal whose marketable value you have determined, by excruciating agony to *him*, to increase. As he lies prostrate, all that is necessary to save him the smallest amount of pain, is to desire the operator with his left hand to close the animal's upper nostril, while beneath the lower one he places a quarter-of-a-pint tin pot containing a sponge, in which is gradually dropped from a little phial chloroform sufficient to deprive him of sensation, which can be readily tested by the occasional prick of a pin."

Why should the neglect of a veterinary surgeon to use chloroform not be treated as an offence against the Act for Preventing Cruelty to Animals?

Besides advocating humanity to the horse, whether in health or sickness, our author most properly insists on the duty of sparing the poor animal all needless suffering, when at last he is sent to the slaughterer— "*l'équarrisseur*," as he is termed in Paris—and of whose operations he gives an interesting account: " Before me was a mass of about fifty yards of motionless and moving substances. The former were the carcasses of horses, at the furthermost end, in their hides; nearer, just skinned; nearer still, headless; and close to me divided into limbs. Among this mass of skulls, bones, limbs, and dull flabby skins, stooping and standing in various attitudes, were the men who were performing these various operations. In a portion of the yard, about fifty yards off, I found standing, tied up to a strong rail, the three horses next to be slaughtered. What were their disorders, of lungs or limbs, whether they were broken-winded or incurably lame, were facts I did not care to investigate; but there is something so revolting in the idea of allowing a poor horse—our willing servant-of-all-work—to suffer in his last moments from the pangs of hunger, that I was glad when a single glance at their flanks showed me that they were full of food. They are not allowed to be kept alive above twenty-four hours. During the time they are alive, horses, cows, and bullocks receive one *botte* of hay per day; asses and mules half a *botte*.

"A few yards off was a large heap of horses' feet, and as most of them had shoes on, I inquired the reason. '*Ah!*' said the man, very gravely, '*c'est qu'ils ont appartenu à des personnes qui ne s'amusent pas à les déferrer:*' 'Ah, it is because they belonged to people who did not care about taking the shoes off.' He then conducted me to a covered building, where the bodies of the horses are boiled, and in which are steam-presses to extract horse-oil, after which is made Prussian blue, the residue being sold as manure."

The visit to the slaughter-house for pigs is perhaps the most comic chapter in the 'Faggot of French Sticks;' but as we treat only of the horse at present, we must entreat our readers to peruse it for themselves. We shall only state the Parisian rule—" No man has a right to kill a pig in Paris." How different "in England, where anybody, in one's little village, from the worthy rector at the top of the hill down to the little ale-house-keeper at the bottom, kills a pig! The animal, who has no idea of 'letting concealment, like a worm in the bud, prey on his damask cheek,' invariably explains, *seriatim*, to every person in the parish—Dissenters and all—not only the transaction, but every circumstance relating to it;" and so continues an account of it so ludicrous that we cannot copy it for laughing.

But we must desist from further comment on what Sir Francis Head has to communicate on the multifarious matters stored up in his vigorous and well-furnished intellect. All his works are amusing and instructive; and in the latest of them every owner of a horse may learn much of importance alike to himself and his steed.

We cannot conclude without expressing an earnest wish that all having horses shall give their servants ' Horse-Taming,' by J. S. Rarey, as what purports to be an unabridged edition has been published at Edinburgh, adorned with a portrait ugly enough to frighten any horse, and sold for a penny. We believe that to subdue a vicious and powerful horse, possessed by an evil spirit such as seemed to have entered into " Cruiser," there is need of more pluck, coolness, and strength than falls to the lot of most men. We do not, therefore, anticipate that every owner of a horse may call forth his good and subdue his evil qualities with the success of Mr Rarey; but every reader of his work may learn how to control himself, and thus to acquire the mastery of his horse.

It must be for the benefit of the horse and his rider that the following paragraph be suspended in every stable : " Almost every wrong act the horse commits is from mismanagement, fear, or excitement : one harsh word will so excite a nervous horse as to increase his pulse ten beats in a minute. When we remember that we are dealing with dumb brutes, and reflect how difficult it must be for them to understand our motions, signs, and language, we should never get out of patience with them because they don't understand us, or wonder at their doing things wrong. With all our intellect, if we were placed in the horse's situation, it would be difficult for us to understand the driving of some foreigner, of foreign ways and foreign language. We should always recollect that our ways and language are just as foreign and unknown to the horse as any language in the world is to us, and should try to practise what we could understand were we the horse, endeavouring by more simple means to work on his understanding rather than on the different parts of his body."

THE ARAB HORSE OF AFRICA.*

WE are about to give our readers some new notions regarding the origin and treatment of the noblest of our domesticated animals, the horse. Our information is derived from a singular book, the most remarkable portions of which are furnished by the Emir Abd-el-Kader, whose protracted resistance to the French in Algiers, followed by long captivity in France, excited such compassion, and whose release, opposed by politicians for reasons of public security, was at length effected by personal appeals to the generosity of the Emperor.

The book is not only interesting in itself as a valuable contribution to the natural history of the horse, it also possesses a peculiar charm to the philosophical student of human nature. Its main subject, no doubt, is the *genus equus;* but, quite unintentionally on the part of the French general and the Arab chief, the *genus homo* stands out prominently, and under aspects so diversified as to afford a most striking exhibition of the many-sidedness of the human being as modified by climate and religion—those influences which so powerfully affect his physical condition and the extent of his mental development.

* 'The Horses of the Sahara, and the Manners of the Desert.' By E. Daumas, General of Division commanding at Bordeaux, Senator, &c. &c. With Commentaries by the Emir Abd-el-Kader. Translated from the French by James Hutton. W. M. H. Allen & Co., London, 1863.

The brave, temperate, half-religious, half-fanatical Man of the Desert, and his chief friend the horse, are brought into sharp contrast with the equally brave, but the much less self-denied, and the much less religious Man of Europe, in the person of a distinguished French general, familiar with the civilisation and learning of his own intelligent nation, and patriotically seeking to turn to its advantage the valuable knowledge acquired during a long residence in Algeria.

Our readers will readily understand the sort of contrast afforded by the respective views of persons so differently trained as a French officer and an Arab chief. Their mutual relations, moreover, have in them something touching. The exiled Arab is politely requested, by an agent of the power which had crushed him, to furnish information regarding the most valued animal of the desert; the object being to supplement works in which General Daumas had served the interests of France by throwing light upon important questions of war, commerce, and government. The Emir, complimenting him on his thirst for knowledge, replies —" You ask me for information as to the origin of the Arab horse. You are like unto a fissure in a land dried up by the sun, and which no amount of rain, however abundant, will ever be able to satisfy. Nevertheless, to quench, if possible, your thirst (for knowledge), I will this time go back to the very head of the fountain. The stream there is always purest and freshest. Know, then, that among us it is admitted that Allah created the horse out of the wind, as he created Adam out of mud. Several prophets—peace be with them!—have proclaimed what follows : When Allah willed to create the horse, he said to the south wind, ' I will that a creature should proceed from thee—condense thyself!' and the wind condensed itself. Then came the angel Gabriel, and he took a handful of this matter and presented it to Allah, who formed of it a dark bay, or a dark-chestnut horse, saying, ' I have called thee horse

—I have created thee Arab—I have attached good fortune to the hair that falls between thy eyes; thou shalt be the lord of all other animals; men shall follow thee whithersoever thou goest. Good for pursuit as for flight, thou shalt fly without wings.'"

And then, by a most curious process of reasoning, the Emir proceeds to demonstrate that Allah created the horse before Adam, and the horse before the mare! His proof of the latter fact is accompanied by an assertion of such practical importance that it deserves notice. "My proof is that the male is more noble than the female, and he is, besides, more vigorous and patient. Though they are both of the very same species, the one is more impassioned than the other, and the divine power is wont to create the stronger of the two the first. What the horse most yearns after is the combat and the race. He is also preferable to the mare for the purposes of war, because he is more fleet and patient of fatigue, and because he shares his rider's emotions of hatred or tenderness. It is not so with the mare. Let a horse and a mare receive exactly the same sort of wound, and one that is sure to be fatal, the horse will bear up against it until he has succeeded in carrying his master far from the field of battle; while the mare, on the contrary, will sink at once upon the spot, without any force of resistance. There is not a doubt on the subject; it is a fact known by proof among the Arabs. I have seen frequent instances of it in our combats, and have experienced it myself."

It is this singular mixture of fanciful reasoning with practical knowledge that makes the Emir's commentaries so worthy of observation. Dismiss his "proof" of the horse being created before the mare, cavalry officers will yet do well to remember the positive assertion of the gallant Emir as to the greater power of endurance characteristic of the horse; and it is equally obvious that it ought not to be forgotten by the gentleman following the hounds, seeing that the integrity of his neck so often depends upon the pluck and bottom of his steed; and it is also to the interest of the agriculturist that he should bear in mind the greater vigour ascribed to the horse.

In France, assuredly, the fact appears to be considered deserving of attention; for, in the old days of the diligence, who does not remember the neighing of the stallions by which it was generally drawn? and we have a vivid recollection of the number of horses in the cavalry regiments of France.

But while the book abounds with striking contrasts between the horse and his rider in Europe and in the African desert, the greatest novelty to many a reader is the political and religious aspect of the horse among Mussulmans. We prize the noble creature for his physical qualities, as well as for his mental endowments. We know his worth as our ally in peace and in war; and his social instincts induce us to be with him on terms of friendliness. The gift of a valuable horse is deemed worthy of the acceptance of the mightiest kings, in symbol of homage or of friendship. And thus the Queen of Great Britain has from time to time received presents of their finest horses from the chief rulers among the nations. When, in 1860, the Emperor and Empress of the French visited Algiers, the various tribes, whose long resistance had yielded to the persevering assaults of the French, were assembled to do homage to the Emperor, the chiefs, clad in their richest dresses, alighted from their steeds, and advanced in a body to present the richly caparisoned *horse of homage.*

Though we know all this, and are aware that the British is the most equestrian nation in the world, most of us will be surprised to learn that, in the faith of Islam, the horse is invested with a kind of sacredness which has signally advanced the purposes of Mohammedan ambition. The Jewish lawgiver, with the design of confining the Hebrews to Palestine, and preserving them a peculiar people, expressly forbade the multiplying of horses, or the fetching of them from Egypt; so that, when we read that Solomon had "forty thousand stalls of horses for his chariots, and twelve thousand horsemen," we know that he was transgressing the Mosaic precepts, even as when he took unto him "seven hundred wives and three hundred concubines" (1 Kings xi. 3).

Mohammed, on the contrary, a great military genius, and bent on the aggrandisement of the Arab race, inflamed their passion for fleet steeds and fair women, and adroitly contrived that horses and houris should become at once potent instruments of religious propagandism and political ambition. All nations and governments have regarded the horse as one of the prime elements of their strength and prosperity. The Prophet of Mecca had the sagacity to make him a principal agent in these furious irruptions of the Arabs which have left such permanent effects upon many of the kingdoms of Asia, Africa, and Europe. With the keenest appreciation of his value in war, he adopted the subtlest expedients to make it subservient to his purposes. "In the days of paganism," observes Abd-el-Kader, "they loved the animal from motives of interest, and merely because it procured them glory and wealth; but when the Prophet spoke of it in terms of the highest praise, this instinctive love was transfigured into a religious duty." When, in token of submission by the five tribes of which Arabia then boasted, five magnificent mares were presented to the Prophet, "it is said that Mohammed went forth from his tent to receive the noble animals, and, caressing them with his hand, expressed himself in these words, "Blessed be ye, O daughters of the wind!" Afterward the messenger of Allah said, in addition,—"Whosoever keeps and trains a horse for the cause of Allah is counted among those who give alms day and night, publicly or in secret; he shall have his reward. All his sins shall be remitted, and never shall fear dishonour his heart."

A volume might be filled with phrases from the Koran, the traditionary sayings of the Prophet, and the commentaries upon them, all inculcating the love of horses as a religious duty.

The believing Emir, for prudential reasons probably, appears to be blind to the political results of this religious equestrianism. Not so is General Daumas. With the clearest insight into what Mohammed proposed to himself as a great conqueror, he maintains that it was essential that the horse should be looked upon in the light of a sacred animal, a providential instrument of war, created by the Deity for a special purpose, and of a nobler essence than that of which he fashioned the other animals. As the result has proved, he herein showed himself thoroughly acquainted with the temperament of the people who were to be the instruments of his soaring ambition. This policy had sense in it. To worship sacred bulls as the Egyptians did, and the Hindoos do to this day, is a foul degradation of the human intellect, bringing with it no conceivable benefit by way of compensation. Hence while with Milton we scorn fanatic Egypt—

"Likening his Maker to the grazed ox,"

we see so much method in Mohammed's fanatical propagandism by means of the sacredness he associated with the horse, that we cannot possibly regard him as a moon-struck dreamer. No, verily! with the hoof of the horse he has left his ineffaceable mark upon the human race.

If, then, we desire to be acquainted with the early history of this precious quadruped, to know its capabilities, and be familiar with the treatment of it by the people who love it most and understand it best, we must talk with the Arab in his tent, and wander with him in the desert.

The Arab and *the* Desert, to which we are introduced

by General Daumas and Abd-el-Kader, demand a few words of explanation.

The *man* is the descendant of those colonists of the Arab race who permanently settled in Africa during the sixth and seventh centuries. The *region* into which, eventually, they have been driven is the Sahara or *Great Desert*, strictly so called, seeing that it is the most extensive on the surface of the earth—its area being estimated at 2,500,000 square miles, or two-thirds that of Europe. For hundreds of miles the eye only rests on bare sands in flats and hillocks, or on naked and rocky tracts, destitute of vegetation, and seldom exhibiting any of the forms of animal life. Nevertheless the Sahara contains numerous fertile tracts (*oases*), watered by perennial springs, and containing a numerous population, consisting of two nations of Berber origin (the most ancient inhabitants of North Africa, according to some, and descended from the Phœnicians), but divided into numerous tribes, all Mohammedans. The *fauna* of the Sahara is as deficient as its *flora;* and, notwithstanding the frightful heat during the day, the nights, owing to excessive radiation, are so cold that ice is frequently formed.

In this sterile region there are patriots and poets and warriors, all ready to die for fatherland, and calling on the nations of the earth to believe that they are the people favoured of Allah!

Our learned Emir can wield the pen as well as the sword; and aspiring, mayhap, to be the poet-laureate of his tribe, he indites a poem in which we are told—

> "Two things are beautiful in this world,
> Beautiful verses and beautiful tents."

Addressing him who condemns the love of the *Bedoui* for his boundless horizons, our poet asks—

> "Is it for their lightness that thou findest fault with our tents?
> Hast thou no word of praise, but for houses of wood and stone?

If thou knewest the secrets of the desert, thou wouldst think
like me;
But thou art ignorant, and ignorance is the mother of evil."

And then, after a description of the natural features of the Sahara, and the stirring incidents of desert life, comes the proud boast—

" We are kings. There is none to be compared with us.
Is it life to undergo humiliation?
We suffer not the insults of the unjust. We leave him and
his land;
True happiness is in wandering life."

We are very willing to believe it; and seeing that General Daumas declares that there is a striking resemblance between the horseman of the Sahara and the knight of the Middle Ages, we are surprised that many more of our wealthy British travellers do not visit Algiers, with the intention of making a brief sojourn among the singular people inhabiting the not distant Sahara. Our invalids begin to appreciate the charming climate of Algiers, where the seasons glide into each other imperceptibly, the range of the barometer being only from $29\frac{1}{10}$ inches to $30\frac{4}{10}$ inches—the whole cycle of the weather's changes indicated within the range of $1\frac{3}{10}$ inches! What a contrast to the turbulent mutability of a climate like ours! We experience what the Germans call a *sehensucht* for a land where the trees bud in February and the fruit is ripe in May, which also is harvest-time.

And should we ever enjoy the equable temperature of the Barbary States, and take a peep at the genuine Arab chief in the Sahara, we shall doubtless experience his proverbial hospitality; though, by the way, we are startled to find among the schedule of his effects only "three wooden platters for strangers to eat from." We have no notion what they are like, but comfort ourselves with the belief that they must be tolerably big, inasmuch as they are priced 13s. 6d.—rather a large sum in the desert, where coin is so scarce. Should our wife accompany us, it is manifest that the loss of any of her multifarious raiment will not be readily replaced; for the four wives of our desert host and the two wives of his two sons have only, General Daumas warns us, one haick a-piece, one pair morocco-leather boots, embroidered (price 4s. 6d. *each*—we like to be precise when writing for the ladies). *Chemises* and *chemisettes* being nowhere in the inventory, we fear our "second self and mysterious double" will not at once take into her loving arms her Arab sisters, even though assured that the lack of clean linen is by them supposed to be more than compensated by their being each the possessors of two pairs silver ear-rings set in coral.

As to our creature-comforts, we are in hopes that they will not be neglected, for our Arab host's property is estimated at £5121. He will, moreover, have time to make himself agreeable, for all he has to do is to attend the meetings of his tribe, ride about, look after his flocks, and pray.

It is quite a mistake to fancy that in the desert every man has a horse, and that walking is at a discount. The poor Arabs are astonishing pedestrians, and, as special messengers, think themselves well paid with four francs for going sixty leagues. In the desert such a messenger travels day and night, and sleeps only two hours in the twenty-four. When he lies down he fastens to his foot a piece of cord of a certain length, to which he sets fire; and just as it is nearly burnt out the heat awakes him. General Daumas relates that one of these men, the bearer of an important message, travelled about 120 miles in sixteen hours, eating during the journey only a few dates, and drinking about three and a half pints of water. So that the British pedestrian through the Sahara, if such a personage shall appear, bent on "astonishing the natives," will require to step out briskly in order to accomplish his amiable project.

As our compatriots, when in foreign parts, are accused of "coming it grand," and of trusting to the imposing effect of their sturdy *physique*, we beg quietly to give them the hint that the Arabs laugh at braggadocio airs, and are not at all impressed by lofty stature and bodily strength. Their esteem is bestowed on activity, address, and courage. Looking at a burly fellow whose praise is being sung, they may be heard whispering, "What to us is the stature or the strength? let us see the heart. After all, it may only be the skin of a lion on the back of a cow."

But we must not dwell on the manners of the desert—we especially wish to tell about its horses.

The horses of the Sahara are of especial interest, not only on account of their intrinsic merits, but also because of the comparative accessibility of their habitat. If to Europe the infusion of Arab blood has been an undeniable improvement in the breed of our horses—if it be still acknowledged that the thorough-bred horse of Arabia is a most desirable acquisition—the question immediately occurs, Where may he be most readily procured? The answer must be, From North Africa—the Barbary States, Algeria—regions to us much more accessible than the desert of Arabia or our empire in India, where the breeding of horses is systematically prosecuted by Government.

And to repair to these regions we have the additional inducement derivable from the fact emphatically asserted by Abd-el-Kader, and assented to by General Daumas, that the horse of the Sahara—the Barb—is the veritable horse of the Orient, identical with the horse of Arabia, brought into Africa by the ancient race of the Berbers.

It is through the French in Algiers that we may chiefly hope for supplies of this valuable animal; but not even through them, without infinite trouble, and very possibly only of an inferior type, and from an Arab compelled by poverty to part with an animal which is his own friend, and the friend of all his family, petted and prized alike by old and young.

Allah has said—"The horse shall be cherished by all my servants, and none will I place on his back save those who know me and worship me." The Mussulman princes have availed themselves of this politico-religious dogma to prohibit the sale of Arab horses to Christians under pain of sin and damnation. The Arab's love of money cannot be gratified by the sale of a really valuable horse; his whole tribe would resist it. And so, if the repute of the Arab steed among us be not so high as once it was, the explanation is, that the horses and mares of the highest order are never parted with to foreigners at any price. In a recent article in the 'Edinburgh Review,' entitled "English Horses," the writer not only asserts that in India the Arab is confessedly inferior to the English racer, but thinks it probable that, owing to the intercourse between Arabia and India, and the high prices for horses at Calcutta and Bombay, the very best horses that Arabia produces are to be procured in India.

But, in opposition to this, we repeat that religion and policy require of the Arab chiefs that they shall exclude infidels from the possession of their noble breeds of horses. When at the height of his power, Abd-el-Kader inflicted death without mercy on every believer convicted of having sold a horse to a Christian. In Morocco the exportation of this animal, so valuable as an instrument of war, is hampered with such restrictions that the permission to take it out of the country is altogether illusory. At Tunis the same reluctance only yields to the imperious necessities of policy; and in like manner at Tripoli, in Egypt, at Constantinople —in short, in all Mussulman states.

General Daumas asserts—"I know for a fact that, in certain Mussulman countries, in the list of obligatory presents for a Christian personage, the donor wrote down, 'a jade for the Christian.'"

To French *politesse, finesse*, or lightfingeredness, or to French success in war among the tribes of the desert, must we look, then, for pure specimens of the horse, which the Arab identifies with himself and his religion.

"The love of the horse," General Daumas writes, "has passed into the Arab blood. That noble animal is the friend and comrade of the chief of the tent. He is one of the servants of the family. His habits, his requirements, are made an object of study. He is the burden of their songs, the favourite topic of conversation. It is thus that the Arabs acquire that knowledge of horse-flesh which we are so astonished to meet with in the humblest horseman of a desert tribe."

In regard to their belief that in intelligence the horse approaches the nearest to man, the General asks, "May it not be that the Arabs, by living on such intimate terms with the horse, have succeeded in developing faculties the very existence of which is unknown to us, who accord to that animal only the instinct of memory?"

At all events, it is manifest that Arab notions regarding horses are worthy of attention, and that special respect is due to the opinions of Abd-el-Kader, because of his exalted rank in Mussulman society, and his science and skill as a horseman. As we have already said, he insists on the fact that the horse of Arabia, and the Barb or horse of North Africa, are identical. Moreover, the Barbary horse, so far from degenerating from the Arab, is actually his superior, and is the very perfection of a war-horse. The French in Algiers have good reason for being of the same opinion. A *chasseur d'Afrique*, setting out on an expedition fully armed, takes with him coffee, sugar, beans, rice, pressed hay, barley, for five days, along with four horse-shoes, and several things besides, so that the total weight carried by his horse is twenty-five stone; whereas the English light dragoon, accoutred in marching order, weighs nearly nineteen stone. So that well may General Daumas exclaim—"A horse that, in a country often rough and difficult, marches, gallops, ascends, descends, endures unparalleled privations, and goes through a campaign with spirit, with such a weight on his back, is he, or is he not, a war-horse?"

If the Barb should be welcomed by our cavalry regiments, he is also possessed of such speed as to make him an acquisition to our racing studs. A Barbary horse, rode by the owner, M. F. de Lesseps, was the winner at the races at Alexandria in 1836, to the surprise of the Viceroy of Egypt, who had been bantering M. Lesseps on the arrival of a horse sent to him from Tunis.

The fame of the Barb was established in England long ago, for Markham, in 1593, writes thus:—

"Next the Turke I place the Barbarie: they are beyond all horses whatsoever for delicacy of shape and proportion; they are swift beyond other forraign horses, and to that use in England we only imploy them; yet are their races only on hard ground, for in soft or deep ground they have neither strength nor delight. Their colours are for the most part gray or flea-bitten." This remark of Markham as to the general colour of the Barb is a proof that the best specimens of the animal were unknown in England.

"Allah created the horse *koummite*—red mixed with black—that is, dark bay or dark chestnut." This, then, is the colour preferred by the Arabs, and General Daumas agrees with them in this preference. Mohammed and Moussa, the celebrated conquerors of Africa and Spain, must, he argues, have had thorough knowledge of the superior value of the chestnut-coloured horse; and their opinion being that of all the Arabs likewise, it is entitled to respect.

"If," he observes, "it be true that those whose coat is red shaded with black are endowed with superior speed, are we not justified in inferring that such was the uniform colour, such the natural qualities, of the sires of the race? I submit, with all humility, these observations to men of science."

"The Emir, moreover, assures us that it is ascertained by the Arabs that horses change colour according to the soil on which they are bred. Is it not possible that, under the influence of an atmosphere more or less light, of water more or less fresh, of a nurture more or less rich, according as the soil on which it is raised is more or less impregnated with certain elements, the skin of the horse may be sensibly affected? There is, perchance, in all this a lesson in natural history not to be despised; for if the circumstances in which a horse lives act upon his skin, they must inevitably act also, in the long-run, upon his form and qualities."

To sum up Arab sense and nonsense as to colour: the fleetest of horses is the chestnut; the most enduring, the bay; the most spirited, the black; the most blessed, one with a white forehead. "Flee the piebald like the pestilence, for he is own brother to the cow." The Isabel, with white mane and tail, no chief would condescend to mount such a horse. Some tribes would not allow him to remain with them a single night. The Prophet abhorred a horse that has white marks on all its legs.

"A horse with white feet, his off fore-leg being alone of the colour of his coat, resembles a man who carries him gracefully in walking, with the sleeves of his coat floating in the air." But to this *dictum* of the Emir we oppose that of the very clever author of 'Adventures of a Gentleman in Search of a Horse:'—

"A dark hoof is preferable to a white one; the latter is more porous in its structure, and more liable to become dry and brittle. This is easily demonstrated by soaking two hoofs of opposite colour and equal weight in water. The white hoof will become heavier than the other when saturated, and will become dry again far sooner. It is also quite notorious among farriers that when a horse is lame, having one foot white and the other black, the disease is generally found in the white foot. So common is this prepossession against white feet, that I have known instances of them being stained by chaunters; but while I admit that a preference is due to the dark hoof, I cannot say that I would reject him for the want of it."

We must allow our readers to decide whether, in the matter of the colour of a horse's foot, they shall side with English blacklegs, *alias* chaunters, or with the Emir of the African Desert.

The late Lord Chancellor Campbell wrote a book demonstrating that Shakespeare was learned in the law; and, more recently, Bishop Wordsworth has done the like to prove that he was a good Christian and an orthodox divine: a horse-dealer may with equal truth make it clear that he was very knowing in horse-flesh; for has he not drawn this portrait of a horse as he ought to be?—

> "Round-hoofed, short-jointed, fetlocks sharp and long,
> Broad breast, full eye, small head, and nostril wide,
> High crest, short ears, straight legs and passing strong,
> Thin mane, thick tail, broad buttock, tender hide."

It is curious to contrast Shakespeare's *beau ideal* of a horse with that of an Arab of the desert:—

"A thorough-bred horse," says Abd-el-Kader, "is one that has three things long, three things short, three things broad, and three things clean. The three things long are the ears, the neck, and the fore-legs; the three things short are the dock, the hind-legs, and the back; the three things broad are the forehead, the chest, and the croup; the three things clean are the skin, the eyes, and the hoof. He ought to have the withers high and the flanks hollow, and without any superfluous flesh. The tail should be well furnished at the root, so that it may cover the space between the thighs; the tail is like unto the veil of a bride. The eye of a horse should be turned as if trying to look at its nose, like a man who squints. The ears resemble those of an antelope when startled in the midst of her herd. The nostrils wide: each of his nostrils resembles the den of a lion; the wind rushes out of it when he is panting. The cavities in the interior of the nostrils ought to be entirely black. If they be partly black and partly white, the horse is only of moderate value."

We pray our readers to note that the Arabs never mutilate their horses, as too many of us still do. Their ears and tails are left as nature made them, on the propriety of which primitive treatment we cannot do better than quote from that most wise and witty book, 'The Horse and his Rider,' by Sir Francis B. Head:—

"About forty years ago it was the general custom to dock the tails of all hunters, covert hacks, and waggon-horses so close that nothing remained of this picturesque beautiful ornament of nature but an ugly stiff stump, very little longer than the human thumb, which, especially in the summer-time, was seen continually wagging to the right and left, in impotent attempts to brush off a hungry fly biting the skin more than a yard off. At about the same period an officer in our army took to the Cape of Good Hope a gentle, beautiful, thorough-bred mare, which, to his astonishment, the natives seemed exceedingly unwilling to approach. The reason was that her ears had been cropped; and as among themselves that punishment was inflicted for crimes, they were induced to infer that the handsome mutilated animal had suffered from a similar cause—in fact, that she was *vicious*."

But if a Mussulman think it a profanation to dock the ears and tail of a horse, what would he say of the ignorant barbarity of too many Christians, who actually inflict blindness on the horse by extirpating *the haw*—that curious appendage to the inner angle of the eye of a horse, which consists of a dark membrane, whose rapid transit over the eye, apparently at the will of the animal, cleans the eyeball of dust or other particles?

When this membrane is slightly inflamed, and thus more prominent than usual, most country farriers cut it off as a diseased excrescence. And so wide is the delusion that in the Encyclopædia of Rees, under the article *haw*, this important membrane is described as a diseased tumour in the eye, and instructions are given for removing it!

We beseech all farriers and horse-fanciers to remember that it is shameful to be ignorant of the physical formation of the most prized of domestic animals. "Sidi Aomar, the companion of the Prophet, hath said, 'Love horses; tend them well, for they are worthy of your tenderness. Treat them like your own children; nourish them like friends of the family; clothe them with care. For the love of Allah, do not neglect to do this, or you will repent of it *in this house and in the next.*'"

The races of the horse most esteemed by the Saharenes are three; of these that of the Hâymour, the foal of the onager or wild ass—so called because a celebrated mare, abandoned as being seriously hurt, is believed to have been covered by a wild ass (*hamar el ouâhhch*)—is the most remarkable. "Whoever," says the Emir, "has seen the horses of that breed will not for a moment question the truth of the tale, for their resemblance to the zebra strikes every eye." This is a curious illustration of the value of the French experiments in hybridising, explained in our article on the acclimatisation of animals.

The Arabs affirm that the best age for reproduction is from four to twelve years as regards the mare, and from six to fourteen as regards the horse. They agree with British breeders in believing that the foal receives more of its characteristic qualities from the horse than from the mare; hence their proverb, "The foal follows the stallion," who also is believed to transmit his moral qualities. "The noble horse," said the Arabs of old, "has no vice."

Stallions, however, are rare in the desert, and belong

only to the principal chiefs, who can afford to have them properly looked after, as it would be dangerous to turn them loose on the grazing-grounds. The Arabs generally prefer mares, because, in the time of war, they do not betray their movements by neighing like the horse, and are, moreover, more easily kept. In the value of the mare as a breeder they have an additional reason for this preference. Hence they exclaim, "A mare that produces a mare is the head of riches;" and with perfect truth, seeing that from three to four thousand pounds have been received for the offspring of a single mare.

The foaling exhibits certain Arab peculiarities. The instant the foal is dropped it is taken up in the arms of one of the bystanders, who carries it up and down in his arms for a considerable time, in the midst of almost inconceivable din, purposely made under the belief that, after such a horrible uproar, it will never afterwards be frightened at anything! This lesson over, the master of the tent places the right dug of the mare in the foal's mouth, and exclaims, " May Allah bring us good fortune, health, and abundance!" All his friends respond, " Amen! may Allah bless thee! He has sent thee another child."

Besides teaching the foal to suck its dam, it is speedily taught to drink camel's and ewe's milk, because it can thus be left in the tent when the mare is again put to work; and because, in default of water, it will in after life be satisfied with milk instead, not only as drink, but as food, should barley run short.

The weaning is generally in the sixth or seventh month, when the women take possession of the foal, saying, " It belongs to us now; it is an orphan, but we will make its life as pleasant as possible." And they do. The women and the children play with it and pet it, feeding it with a sort of semolina made with wheaten flour, as well as with bread, milk, and dates; and to their gentle handling the Arab horse is doubtless indebted for his admirable docility.

We used to fancy that Mohammed had a low idea of the fair sex, but, with our Emir to instruct us, we now know better. The messenger of God has said, "The greatest of blessings is an intelligent woman or a prolific mare." But the Prophet was so *hippolatrous* that we suspect he valued *houris* in proportion to their being lovers of horses.

It is in the early education of the colt, and in soon accustoming him to fatigue, that the Arabs specially differ from us. "Every horse inured to fatigue brings good fortune." At the age of eighteen to twenty months the colt is mounted by a child, who takes him to water, goes in search of grass, or leads him to pasture. The boy and the colt are thus simultaneously educated; the one to be a horseman, the other to fear nothing, and obey the will of his owner.

The colt is also now accustomed to be shackled with clogs, of which mode of shackling General Daumas highly approves:—

"With it one never hears of a horse breaking loose —a misadventure that causes such confusion in a bivouac, drives horsemen to despair, and is the source of a thousand accidents. The Arabs are loud in their abuse of our mode of tying up horses with a longe. They affirm that, in addition to the accidents it may occasion, it has the great inconvenience of not allowing the animal to lie down; whereas with clogs a horse protrudes his head and neck, and when he wants to sleep, places himself exactly in the position of a greyhound basking in the sun. Besides, a great many stable vices disappear when they are used: the animal can neither entangle itself in the halter, nor slip it, nor get into the manger, nor lie down beneath it, nor scratch the earth with its foot, nor rub against the manger, nor contract any bad habits of the kind: an indisputable advantage so far."

At the age of from twenty-four to twenty-seven months the colt is cautiously saddled and bridled—the bit being for the first few days covered with undressed wool. Wealthy owners, before allowing their colts to be mounted by a grown man, sometimes have them led up and down gently for a fortnight with a pack-saddle on their backs, supporting two baskets filled with sand.

"Suppose the colt now to have completed two years and a half, his vertebral column has acquired strength—the clogs, the saddle, and the bridle are familiar to him. A cavalier mounts on his back. The animal is certainly very young, but he will be ridden only at a walking pace, and his bit will be a very easy one. The main point is to accustom him to obedience. The owner—without spurs, and holding only a light cane, which he uses as little as possible—rides him to the market, or to visit his friends, his flocks and pastures, and attends to his affairs, without exacting anything more than submission and docility. This he ordinarily obtains by never speaking to him except in a low voice, without passion, and carefully avoiding anything likely to elicit opposition, that must result in a contest from which he might come forth conqueror but at the expense of his horse. Particular importance is attached to keeping the young animal still and quiet for a few minutes before letting him start."

When the colt is about thirty months old he is taught not to break loose from his rider when dismounting, and not even to stir from the spot where the bridle has been passed over his head and allowed to drag on the ground. This lesson, so important to Arabs, and which all horses should learn, is taught thus:—A slave stands beside the colt and puts his foot on the bridle whenever the animal is about to go off, and thus gives a disagreeable shock to the bars of his mouth. After a few days of this exercise he will stand stock-still at the place where he has been left, and to which he possibly fancies that he has been fastened. What would a farmer think if he saw a man go into the midst of a horse-market, pass the bridle over his horse's neck, let it fall to the ground, put a stone upon it, and then proceed for a couple of hours to transact his business? And yet this sight, the result of the simple process we have described, may be constantly seen among the Arabs. Kneeling is the perfection of the Arab horse's education. The colt is trained to it by tickling him on the coronet, pinching him on the legs, and forcing him to bend the knee. After such a training—for which, however, all horses are not fit—his rider has only to clear his feet of the stirrups, stretch his legs forward, turn out the points of his toes, touch with his long spurs the animal's fore arm, and then, as his piece is fired at the marriage-feasts and other rejoicings, his horse will kneel down amid the applause of the young maidens, piercing the air with joyful acclamations.

Apropos to kneeling to young maidens, a very learned man once confidentially asked us if, when making love, he must kneel to his fair one? As we said No, it cannot be expected that we should advise that our horses shall be trained to treat the ladies as if they were queens. But in the hunting-field, the habit of standing still when desired would be a truly useful habit of the horse, for which farmers, country doctors, and clergymen would be thankful, when moving about the fields, dismounting at ill-made gates opening with difficulty, and requiring a two-handed tug, or tarrying at cottage doors where nobody can be got to hold the horse, the goodman being absent, the wife sick, and the children all literally weans (wee anes).

The power of intelligence and gentleness in developing the usefulness of the horse to man is strikingly seen when in the bivouac. The Arab horseman sleeps with his head resting on his horse's shoulder; an arrangement furnishing an easy pillow, and rendering the theft of the horse a difficult achievement.

But it must not be supposed that the Arabs are all gentleness. A horse needing chastisement gets it by means of tremendous spurs, with which the rider draws long bloody wheals along the animal's belly and flanks, which inspire him with such terror that he becomes as tame as a lamb, and will track out his master like a dog.

They have a proverb, "The horseman makes the horse, as the husband makes the wife;" and therefore every Arab trains his own horse. Here, we opine, is a hint worth something to every man wishing to be on comfortable terms with his horse—or his wife.

We certainly resent any interference with our conjugal authority; and husbands and wives, upon the whole, jog on pretty well in the matrimonial yoke. But our horses are treated shockingly ill; their education is intrusted to an ignorant groom or a brutal breaker-in, with hardly a single right idea as to the philosophy of equestrian training—which, by the way, is begun and *finished* by the Arabs at a much earlier age than with us. "The horse," they say, "is a labourer; let him, then, be accustomed to it in good time." They universally fatigue him without mercy when two and three years old, but spare him from three to four years of age. They maintain that sustained work at an early age strengthens the chest, muscles, and joints of the colt, at the same time imparting a docility that will never leave him. After these rude trials are over, his constitution should be developed by rest and abundant diet, because now he will show whether he be worth keeping.

Such, also, is the testimony of M. Pétiniaud, who was commissioned by the French Government to travel through Upper Asia to procure horses of pure Oriental blood. And as the treatment of the horse by the Arabs of Asia and Africa is thus shown to be identical, and as these men are confessedly the best horsemen in the world, and thoroughly acquainted with the modes of training the animal to be serviceable, it is not unlikely that British breeders will ere long be brought to see the advantage of putting the colt to salutary work from its earliest age.

Feeding is very carefully attended to, each horse receiving a ration proportioned to age, temperament, and work to be gone through. Being convinced that milk maintains health and strengthens fibre without increasing the fat, which they dislike, the Arabs give their horses ewe's or camel's milk as abundantly as they can. The great article of horse food is barley eaten out of a nose-bag, and barley-straw. At night, during winter, the horses get quantities of alfa grass (*Lygeum Spartium*), which furnishes the material of so much of the paper manufactured in this country. Our learned Emir informs us that the Saharene gives his horse camel's milk to drink, "which has the property of imparting speed, so that a man, if he takes nothing else for a sufficient time, will attain to such a degree of swiftness that he may vie with the camels themselves!" As we have no such milk we cannot try the experiment on man or beast; but there is a point in which we may imitate the usage of the Arabs, and not of them only, but of most Eastern nations—we mean the giving of barley to horses. "When thou hast purchased a horse, study him carefully, and give him barley more and more every day, until thou hast ascertained the quantity demanded by his appetite. A good horseman ought to know the measure of barley suited to his horse as exactly as the measure of powder suited to his gun. The Prophet has said—'Every grain of barley given to your horses shall secure you an indulgence in the other world.'"

"Give barley to your horses; deprive yourselves to give them still more, for Sidi-Hammed-ben-Youssouf has remarked, 'Had I not seen the mare produce the foal, I should have said it was the barley.' He has also said, 'Superior to spurs there is nothing but barley.'"

Here are some maxims of the Emir showing sense and humanity:—

"Never water your horse after having given him barley; it would be the death of the animal.

"Leave not thy horse near others that are eating barley without he has some likewise, for otherwise he will fall ill.

" If, at the bivouac, your horse is so placed that he cannot move out of the wind that is blowing violently into his nostrils, do not hesitate to leave the nose-bag suspended from his nose; you will preserve him from serious mischief."

The nose-bag in such a case acts the part of " Jeffrey's patent respirator," and hinders inflammation of the mucous membrane lining the air-passages leading to the lungs. In the human subject this would be termed bronchitis, which we have just seen produced in a boy, in consequence of his sitting down when overheated, exposed to the wind and having on no nose-bag—*i. e.*, " comforter " or " respirator."

In the same philosophic dread of a strong wind right in one's teeth, the Emir adds,—" Contrive, if possible, to save your horse from facing it; you will spare him various diseases." " When you dismount, think of your horse before thinking of yourself: it is he who has carried you, and is to carry you again." That deserves the thanks of the Society against cruelty to animals. Moreover, the Emir is qualified to be president of an abstinence society, for has he not written?— " The cavalier of truth should eat little, and, above all, drink little: if he cannot endure thirst, he will never make a warrior—he is a mere frog of the marshes."

And all who remember that " in the midst of life we are in death," will approve of this religious maxim of the Arab cavalier—" When thou mountest a horse first pronounce these words, 'In the name of Allah;' the grave of the horseman is always open."

Moreover, those seeking to reform the turf may learn something from the Mussulman law, which not only forbids racing for a wager, but all betting by persons not personally concerned in the race. In fact, if we will only have the modesty to confess that from an

Arab we may learn something worth the knowing, we must acknowledge that both the horse and his rider will be benefited by the diffusion among us of the views imported by General Daumas from the Great Sahara.

We have not touched upon the second part of his work, in which he gives a vivid picture of the manners of the desert, and we are far from having made our readers acquainted with all that is note-worthy in relation to horses—the special subject of our article. We must desist, however, and now conclude with commending the work to the perusal of all interested in anthropology—

"The proper study of mankind is man."

Who can fail to wish to know more of the singular being who, in the conclusion of a 'Chant to his Warhorse,' draws this picture of himself?—

"I am an Arab. I know to command and to combat;
My name protects the feeble and the afflicted;
My flocks are the reserve of the poor,
And the stranger in my tent is named the Welcome One.
The Almighty has loaded me with his gifts,
But time turns upon itself, and turns back,
And if I must drink one day of the two cups of life,
I will show that adversity cannot humiliate my soul.
My virtue shall be resignation,
My fortune the contempt of riches,
My happiness the hope of another life;
And if poverty were to grasp me by the throat,
I would not the less glorify Allah."

ACCLIMATISATION SOCIETIES.*

The geographical distribution of animal life, the means by which it is effected, the limits within which it is confined, the extent to which the different members of the animal kingdom have been or may be diffused over the globe,—these are the interesting questions which nowadays engage much of the attention of individual naturalists, as well as of those societies with whose doings, at home and abroad, we wish the public were better acquainted. As these societies also extend their inquiries to the vegetable world, and are continually making us familiar with the valuable properties of plants, and announcing the steps which are being taken for their introduction into other lands than those to which they are indigenous, they are evidently destined to play an important part in altering the physical conditions of future generations. We are confident that the men of the next century will have at their tables the flesh of beasts, birds, and fish, and also fruits and vege-

---

* ' Bulletin de la Société Impériale Zoologique d'Acclimatation, 1864.'

' Fifth Annual Report of the Society for the Acclimatisation of Animals, Birds, Fishes, Insects, and Vegetables within the United Kingdom, 1865.'

' Acclimatation et Domestication des Animaux Utiles.' Par M. Isidore Geoffroy Saint-Hilaire. Quatrième édition.

tables, of which we know nothing save by report. In the good times coming, through the successful labours of the Acclimatisation Society of Great Britain, Ireland, and the Colonies, we hope to dine upon Chinese lamb and Chinese yams, and to find them agreeable substitutes for British lamb and lettuce; and as such things have appeared at the first annual dinner of this Society, it does not seem too sanguine to hope that we too may partake of such delicacies as Kangaroo ham, Syrian pig, Canadian goose, guan, curassoa, pintail ducks, Honduras turkey, dusky ducks, and leporines.

It must be admitted that there is room for diversifying our customary fare. Without being over-addicted to creature comforts, it is allowable to experience a longing for something less common than the mutton, beef, and pork on which our cooks exert their talents. And when our posterity read of the pains we were at in enlarging the means of public alimentation, and in adorning our country with such noble trees as the Wellingtonia, the Araucaria, and the Deodara, they will not only be grateful, but stimulated, let us hope, to imitate our example, and exert themselves in widening the sphere of human knowledge and enjoyment.

We think too little of the influence of man over animated nature, and of his power of surrounding himself in all climates with the productions and the animals which he requires in order to fulfil his mission to possess and subdue the earth; and, by not reflecting upon what has been already accomplished, we deprive ourselves of the stimulus to hope furnished by the contemplation of the vast results already achieved. We are living upon animals and vegetable productions which in the course of ages have been acclimatised and naturalised slowly, because, in many instances, not designedly, but through the intervention of accident and foreign conquest. It is not pretended that this is the native country of wheat, oats, or barley; of the apple, the pear, or the cherry; of the horse, the ox, the ass, the sheep, or the goat; of the goose, the peacock, the pheasant, the fowl. It is evident, therefore, that, by a benignant arrangement of Providence, certain species of animals and vegetables may be transported from their native seats, and made to accompany man in his cosmopolitan migrations. If it were not so his dominion over the earth would be inconveniently limited. If to his new country the emigrant could transport none of the domestic animals by which he was wont to relieve his labour or support his frame, and none of the vegetable productions with whose beauty and usefulness he was familiar, it is easy to understand how much his happiness would be lessened, and how seriously the progress of civilisation would be retarded. Of course there are limits to man's power over nature : all his science and care can be usefully applied only under certain conditions. A polar bear or a reindeer may be transported to the torrid zone, but they will evidently pine or speedily perish. In accommodating itself to new climatic conditions no animal is so cosmopolitan as man; but it is as the companions of his migrations that our domestic animals are now in countries far distant from those in which they had their origin. As he has come from the east, Asia is the birthplace of by far the greater proportion of those which he has subjected to his dominion : this much is certain. If we inquire into the dates at which they were respectively domesticated, we must be contented with general statements. M. Isidore Saint-Hilaire thus sums up his learned researches :—

Among the forty-seven animals actually possessed by man, we find domesticated—

1. From the remotest antiquity, fourteen animals; namely, eleven mammalia—the dog, the sheep, the goat, the horse, the ass, the ox, the zebu, the pig, two camels, and the cat; two birds—the pigeon and the fowl; and one insect—the silk-worm.

2. From Grecian antiquity, four animals: namely,

three birds—the common pheasant, the peacock, and the guinea-fowl; and one insect—the bee of southern Europe.

3. In Roman antiquity, three animals: namely, the rabbit and the ferret; and one bird—the common duck.

4. In antiquity still, but at an undetermined period, two animals: namely, the buffalo and the common bee.

5. At an undetermined epoch, but most probably for several species corresponding to the middle ages, twelve animals: namely, five mammalia—the yak, the reindeer, the lama, the alpaca, and the guinea-pig; two birds—the swan and the ring turtle-dove; two fishes—the carp and the goldfish; and two insects—the Egyptian bee and the cochenille.

6. At an undetermined but probably modern period, five animals: namely, two mammalia—the arni and the goyal; one bird—the Chinese goose; and two insects —the almond and the ailanthus silk-worm.

7. In the sixteenth century, three animals, all birds—the canary, the turkey, and the Muscovy duck.

8. In the eighteenth century, four animals, all birds—gold, silver, and ring pheasants, and the Canadian goose.

Such is the meagre list to which we of the nineteenth century have made no great addition, unless in the class of birds chiefly reared in zoological gardens, and of mammalia of very recent introduction.

The human race has now existed for an unknown course of ages say some, and assuredly for about six thousand years; but man is still very far from having subjected to his sway the hundred and forty species of animals now known to him.

"And of the forty-seven domesticated species," says M. Saint-Hilaire, "fifteen are wanting in France, thirteen in Europe. Is this a sufficient conquest of nature? Is it enough to have in our court-yards three species so valuable as that of the *gallinaceæ*, and only one of the *rodentia*, so remarkable for its fecundity, the precocity of its development, and the excellence of its flesh.

Among the large herbivorous *mammalia* is it enough to possess only four alimentary species? In the middle of the nineteenth century, and surrounded by the marvels daily springing up in mechanics, physics, and chemistry, we have come to this—that the poor still want meat, and that the richest can only vary the meals on their table by varying the preparation of always the same meats; among the large animals, the flesh of the ox, the sheep, and the pig, the milk of the cow, the goat, and the sheep. This is all! With such facts, can we suppose that our civilisation is at every point advanced as far as possible? In regard to our alimentation, can we reckon ourselves as advanced as in regard to our means of transport and correspondence? Have we done for our health what we have done for our industry? Singular contradiction! which we do not observe because habit renders us familiar with it, but which will some day excite astonishment as the most inexplicable of anomalies. Almost in everything else, progress so rapid that what was yesterday now seems separated from us by ages; and, in this fundamental matter of which we treat, progress so slow, or rather such non-progression, that in regard to the number of our butcher-meat species we are where were the Romans, the Greeks, the ancient Egyptians, and, to sum up everything, where the Chinese themselves long have been!"

It is not easy to account for the arrest which has been so long laid on the domestication and naturalisation of animals. Bearing in mind the fact that animals of warm climates bear transportation to colder regions better than the animals of cold climates endure removal to those which are warmer, it is astonishing that nations inhabiting the temperate quarters of the globe have had so little desire to acquire the useful animals abounding in warmer latitudes. This can hardly be owing to the difficulty of transport, and the consequent expense, which is, no doubt, considerable. Suppose that we were destitute of the horse or the ox; it is incredible that the cost of transferring them from some distant region should deter us from attempting their naturalisation among us. We are therefore inclined to attribute the slow introduction of the different species of animals to ignorance of the fact that many of them will thrive when obliged to live on vegetable matters unlike those to which they have been accustomed. The mode in which natural history has been studied has had much to do with the result which we deplore. Under pretence of being rigorously scientific it did not concern itself with practical utility. But now the useful application of the knowledge acquired by the labours of the traveller and the naturalist is that which causes them to be so generally appreciated. The discovery of a new production, or of a new property in some production already known, is hailed as an acquisition to the general good. When science is thus valued its progress is assured and rapid: the speculations of philosophy become the everyday facts of common life.

The useful application of all knowledge is also immensely promoted by the gregarious tendency of our modern society. This is the age of companies, societies, committees: co-operation gives us zoological gardens, and, last of all, Acclimatisation Societies, whose avowed object is to disseminate the fruits of the tree of knowledge, and plant its seeds wherever there is the prospect that they shall germinate and prosper.

In this rivalry to be first in scattering the bounties of Providence, France has been, and is, conspicuous. Her naturalists have been eminently successful in popularising their science by demonstrating its public utility. Cuvier and Geoffroy Saint-Hilaire have found zealous successors in the distinguished men who conduct the Acclimatisation Society, whose yearly Report is a most valuable repository of all relating to the practical application of whatever is known in natural history. Whoever is in search of information the most recent and trustworthy regarding fish-culture, silk-culture, leech-culture, pearl-culture, the introduction of new animals and vegetable productions, and a host of matters which cannot be readily enumerated, let him consult the Bulletin of this Society.

Moreover, it is exercising a most happy influence in bringing together and fusing the various ranks of the French people themselves. Its roll of three thousand members comprises men of every shade of opinion, men of leisure and men of science, simple citizens and the illustrious of the land, men of yesterday and men of ancient family; so that it is, as it were, a compend of all French intelligence—France in miniature, but France without division, and in perfect union.

The charm which has effected such a marvel is simple. It is the aiming at a common object, and the calling in of science to aid in realising a noble idea. In such an association, originating in love of humanity, nationalities commingle, classes are effaced, prejudices disappear, and enmities are extinguished. And thus, to use the beautiful similitude of M. Drouhyn de Lhuys, like many sailors, setting out from divers distant shores, they arrive at the same port, led by the same guiding star. In order that it may not be supposed that all this is too much *couleur de rose*, we must give some details in explanation of the nature of the Society's operations.

In the Bulletin of the Imperial Society, February 1864, the following extraordinary prizes are offered:—

1. Complete domestication and application to agriculture, or to home work, of the Hémione (*Equus Hemionus*) or Dauw (*E. Burchellii*).—Medal of 1000 francs.

2. Introduction and domestication of the Dromée (Casoar of New Holland, *Dromaius Novæ Hollandiæ*) or the Nanitou (American ostrich).—Medal of 1500 francs.

3. Acclimatisation in Europe or Algeria of a wax-producing insect other than the bee.—Medal of 1000 francs.

4. Creation of new varieties of Chinese Ignamas (*Dioscorea batatas*) superior to those we have, and of more easy cultivation.—Medal of 500 francs.

5. Propagation of the Graux de Mauchamp breed of sheep out of the locality in which it originated (in France or abroad).—Medal of 1000 francs, with 1000 francs offered by M. Davin. Proof must be supplied of the possession of at least 100 animals, born on the property of the owner, and exhibiting the type of the Mauchamp breed in wool and conformation.

6. Introduction and acclimatisation in Martinique of an animal destroying *Bothrops lanceolé* (vulgarly known as the iron-lance viper).—Medal of 1000 francs.

7. Introduction, cultivation, and acclimatisation of Quinquina in Southern Europe, or in a French colony. —Medal of 1500 francs.

8. A cross of the Hémione or its congeners (dauw, zebra, couagga) with a mare.—Medal of 1000 francs.

9. Propagation of the cross of the Hémione and its congeners with the ass.—Medal of 1000 francs.

10. Domestication of the African ostrich in Europe. —Medal of 1500 francs.

11. Domestication of the same bird in Africa.— Medal of 1500 francs.

12. Introduction into France, and reproduction in captivity, of the Honduras turkey (*Meleagris ocellata*).— Medal of 1000 francs.

13. Reproduction in France of the pinnated grouse (*Tetrao Cupido*).—Medal of 1000 francs.

14. Acclimatisation of a new edible fish in the fresh waters of France, Algeria, Martinique, or Guadaloupe, or of edible crustacea in the fresh waters of Algeria.— Medal of 500 francs, to be doubled if the fish be the Gourami.

The importance attached to the acclimatisation of the yak is demonstrated by the offer of prizes of 2500 francs and 2000 francs for four pure yaks; and of nine prizes of from 1800 to 200 francs for crosses betwixt the yak and the cow.

The yak, or ox of Thibet and Tartary, was introduced into France by M. de Montigny in 1854. The little herd consisted of five males and seven females.

They are all small, particularly the cows, which are about the size of the small Brittany breed. Their heads and limbs are stunted, and their bodies proportionally a little longer than the common cow. Their rump is round, and somewhat like that of the horse. Their tail is abundantly furnished with long hairs, but not so stiff as those of the horse. The hair of the body is generally straight or slightly curled, with little gloss, and long like that of the goat. The hair is in Thibet made into a thick strong cloth, admirably suited for agricultural labourers. The milk of the yak is excellent, and, according to chemical analysis, is rich in albumen and caseine. Travellers assert that its flesh is very good: of this the Parisian *savans* were speedily able to speak from experience. One of the young bulls born in France having become blind, the Council of the Acclimatisation Society resolved that it should be fattened for the table. Cooked in a variety of ways, it was pronounced excellent. Here is the report of M. Quatrefages:—
" Its flesh is redder than that of the calf; its fibre is equally fine. It has a peculiar and very good flavour, something betwixt mountain-veal and beef, but somewhat *sui generis*. Its juiciness is perfect. In short, we conclude that beef-steaks and fillets of yak should be superior to like parts of the ox. I do not think that the novelty of this repast has in the least increased our appreciation of it. I am satisfied that the day will come when epicures will thank the society for having acclimatised this new ox, which I shall not the less continue to look upon as the future ox for the poor."

Useful as an industrial and alimentary animal, the yak is not less so as an auxiliary. In steep places its sure-footedness is greater than that of any other animal. It draws, it carries burdens, and is at the same time advantageously employed as a saddle-beast. It trots rapidly enough, and its step is agreeable.

In answer to those who doubt the utility of such an animal, M. Quatrefages remarks:—" Yes, alongside of your perfect races there is as yet no place for the yak. But these races have not always existed. You have fashioned the horse, the ox, the sheep, in conformity with your wants; why may the like not be done with the yak? The day is perhaps not distant when we shall have its breeds for wool, for milk, for butcher-meat. Alongside of your large farms there are many very small properties. Perhaps the yak is destined to become the ox of the man of small capital, as the ass is already the horse of the poor. Its native rusticity, and the little food which it consumes, appear to assign this as its place. It may never inhabit the prairies of Normandy or the fields of Limagne, but on the hillocks of Vosges, on the heights of Cévennes, in the Alps, in the Pyrenees, it will browse on the short grass which pushes through the snow, as it does in its native land. Perhaps, in short, it may not be very serviceable to France; possibly it may chiefly migrate to the north. What of that? It is not the first time that France has at her own cost made experiments useful to others."

These are the very qualities which we should desire in an animal sought to be acclimatised for the special benefit of the Scottish Highlands. Our loftiest mountains will not be too inhospitable as localities for creatures indigenous to the heights of the Himalayas. Their power of adapting themselves to circumstances alien to those of their natural condition is singularly shown by their thriving in a locality so warm and low as that of Paris; so that we may expect that their introduction into this country would be a boon widely and speedily diffused. In six years the herd of twelve introduced by M. de Montigny had increased to thirty-five. They have also evinced great readiness to cross with other animals of like species. The cross between the yak bull and a French cow is a valuable animal. There can be no difficulty in introducing them into this country; and we trust that ere long we shall see the yak browsing on our hill-sides. We invoke the powerful aid of the Highland and Agricultural Society of Scotland in the promotion of an object in such harmony with the design of its institution, and so specially fitted to benefit the Highlands.

To France also we owe the prospect of having our domestic animals increased by the addition of a creature known from the most ancient times for its swiftness, but believed to be untamable, though its flesh was, and is, prized as an article of diet. We refer to the Dzigguetai (*Equs hemionus*), or wild ass of the Book of Job. In its native regions, Cutch, Goojerat, Tartary, Persia, and many parts of Central Asia, this animal lives in troops under the conduct of a leader, whose alarm on the least approach of danger sends them bounding over hills and rocks with a fleetness exceeding that of an Arab horse. It may seem in the last degree improbable that such a creature shall ere long take its place among our light beasts of burden, and our animals for the saddle and the course; and yet, to those knowing how readily it has multiplied in France, nothing appears more likely. One of them, several years ago, had bred at Paris for the fifth time. No precautions have been taken to protect them during severe winters; and the breed, so far from degenerating, is manifestly becoming more vigorous.

There is another consideration weighing more with those seeking the improvement of agriculture. The cross obtained from the male of the dzigguetai and the female ass resulted, in 1844, in the production of an animal of great beauty and vigour. This crossing has been going on ever since; and the hybrids, which are strong and very swift, have in many instances been broken; and four of them, accustomed to harness either singly or together, have formed part of an agricultural exhibition, attracting almost as much notice as the finest breeds of horses.

Here, then, is a new field for our horse-breeders; they have only to cross the Channel in order to obtain a breed of animals whose utility is undoubted.

Under the general term *Llama* are comprehended at least three species of ruminants, with which Europe is destined ere long, we trust, to make acquaintance. Being mountain representatives of the camel, modified in form and habit in conformity with their position, the acquisition of them should be the especial object of those inhabiting the mountain regions of the north. Their native region is the highlands of Chili and Peru, where they abound, both wild and tame. The llama is of great utility as a beast of burden, capable of carrying a load of about 100 pounds, while the wool of the alpaca is very valuable. Their flesh, also, is as much esteemed as that of the sheep. In the introduction of these creatures Spain has taken the lead; but now experiments on a great scale have been undertaken for the introduction of the llama and the alpaca into France and Australia,—the latter country having lately held out large pecuniary inducements which have led to their being transported thither in large numbers. The difficulty of this great enterprise has been augmented very needlessly by the selfish policy of the governments of Peru and Bolivia in forbidding the export of these valuable animals. Mr Ledger, the energetic conductor into Australia of 400 llamas, alpacas, and vicugnas, spent seven years on his expedition,—two preparing for, and five in carrying it out. He led them 1500 miles through the passes of the Andes, amid which he encountered storms of wind and snow, threatening destruction to the herd and its drivers. It is gratifying to learn that the colonial government of Sydney, appreciating the difficulties of the enterprise, raised the promised reward from £10,000 to £12,000, even though, of the 400 animals embarked, Mr Ledger succeeded in landing only 256 in Australia. A grant of £1500 a-year for the maintenance of the herd declared the importance attached to this addition to the industrial resources of the colony.

France has been most persevering, but most unfortunate, in her efforts to introduce the llama and the alpaca.

Fortunately French naturalists are as persevering as they are intelligent and zealous, so that disasters have only made them more resolvedly bent on the accomplishment of their purpose. They argue that as the llama and the alpaca have thriven in places so little above the level of the sea as Paris and the Hague, there is reason to expect a much greater success when these mountain animals are introduced to congenial localities among the Alps and the Pyrenees.

It is with especial satisfaction that we note the enlightened zeal and long-suffering patience of the French acclimatisers. They are not ashamed to acknowledge mistakes and disappointments.

These to us are as valuable as their successes; and as the moral from which we may be benefited has been furnished at their expense, let us be thankful for the frank avowals of discomfiture. As the raising of over-sanguine expectations is sure to be mischievous, because inevitably followed by despondency and relaxation of effort, let us ingenuously confess that recent anticipations as to the beneficial introduction into Europe of the llama and alpaca require to be modified. The achievement appears to be neither so easy nor so desirable as was imagined.

Such, at least, is the impression made on us by the perusal of 'The Wool-producing Animals of the Andes, and their Acclimatisation in Europe,' by M. Émile Colpaert, sent on a scientific mission to South America by the Minister of Public Instruction.

M. Colpaert declares that his personal knowledge, as well as all information acquired from natives of Peru, forbid him to share in the conviction that the ruminants of the Cordilleras may speedily be seen grazing like sheep in the pastures of Europe. He positively asserts, even with regard to the llama, that all experience demonstrates that, both individually and in its progeny, it degenerates when not inhabiting the heights of the Andes. He assigns a mechanical reason for this degeneracy, which may possibly be guarded against, with the help of the dentist! Like other ruminants, it has incisor teeth only in the lower jaw. Of these it has only four, which, by a wise provision of nature, are bent slightly forward so as to enable it readily to seize the almost stemless *ycho*, the nutritive sap of which is confined to the root. The incisors being constantly rubbed against the hard soil are kept in such a condition as suits the function of the mouth; but when the llama is fed on cut grass, or on lucerne, they become so elongated as to rub off the skin of the upper lip. Being unable to bring its jaws together without pain, the creature ceases to eat enough; what it does eat is not rightly digested; death ensues.

M. Colpaert became acquainted with this important fact in consequence of being on such friendly terms with some of his suffering llamas that they allowed him to examine their mouths without discharging in his face the abominably offensive saliva which they spit out when annoyed. The discovery of the woeful state of their upper jaws convinced him that the only way to save their lives was at once to despatch them to an *ycho*-producing locality.

We cannot doubt that in this fact, so specially noted by M. Colpaert, we have a chief cause of that mortality which has hitherto characterised every attempt to acclimatise the llama. He also points out that the introduction of this creature into Europe would be a mistake, because its scanty wool is quite mediocre in quality, and not to be compared to that of the Merino sheep. He also depreciates its usefulness as a beast of burden, maintaining that those writers are romancing who speak as if it readily allowed itself to be mounted by anybody; the fact being that it cannot carry a weight much above a hundred pounds, and that the Indians, when travelling, mount their young children only on llamas of approved docility and sobriety of demeanour. He also mentions the singular fact that the llama commits suicide. When overloaded and overdriven it lies down, and when its brutal tormentors exhaust all the resources of cruelty, the unhappy animal at last kills itself by suddenly dashing its head against the earth. Why, then, be at such expense in removing from its native haunts a creature whose wool is inferior to that of our sheep, and which as a beast of burden is so inferior to the ass and the mule? The difficulties, doubts, and queries thus suggested are too many and too grave to be disregarded; and we doubt not that they will induce many to regard the introduction of the llama as chiefly one of curiosity to students of natural history. But the matter may be looked on from another point of view, to which we shall advert in a little.

M. Colpaert, while fully admitting the great value of the alpaca, is still less sanguine that its fleece will enrich the agricultural and textile industry of France. What herb have we to substitute for the indispensable *sora?* What mountain in Europe shall take the place of the Cordilleras? M. Colpaert, while hoping that his fears are chimerical, confesses his inability to reply to such questions.

All that the French acclimatisationists can venture to say is, that the results obtained from some of the animals introduced into France by M. Roehn appear fitted to inspire greater hope of the acclimatisation of the alpaca. They comfort themselves, moreover, with the reflection that the introduction of the Merino sheep into France was only effected after the loss of innumerable animals and the lapse of a century. But we are sorry to observe that their inquiries at the different zoological societies in Europe confirm all that is advanced by M. Colpaert as to the difficulty of removing the alpaca from its native soil. In June 1864 there were in Europe 104 llamas, but only 4 alpacas; namely, 2 at London, and 2 at Paris.

On the principle of hearing both sides of the question we shall contrast the experience of M. Colpaert with that of Mr Ledger, who, after frightful dangers and losses, has succeeded in introducing llamas and alpacas into Australia.

Of the 336 animals which Mr Ledger embarked for Sydney he lost 80 before his arrival in November 1858, notwithstanding the assistance rendered during the passage by thirteen Indians hired for the express purpose of attending to the precious cargo.

On the whole, notwithstanding some mishaps, due mainly to the great drought of 1862-63, and to the alpacas being allowed to breed too soon after their arrival, Mr Ledger maintains that his great experiment is far from being unsuccessful. In February 1864 there were in the colony 328 animals; the wool had improved, and was acknowledged to be very fine. Mr Ledger makes several interesting statements in his paper read to the Acclimatisation Society at Sydney.

"When, fifty years ago, Merinos were introduced into the colony, if any one had spoken of the probability that Australia should one day be the rival of Spain in the production of these fine wools, he would not have been believed. Shall I then be laughed at when predicting that in thirty years Australia will produce three million pounds of alpaca wool?". The hopes of Mr Ledger may be over-sanguine, but it must be confessed that his doings justify distrust of the far more discouraging report of M. Colpaert, so that we should not despair of localities suited to the alpaca being discovered in some of the diversified climates of the British possessions.

The like modified statement, which we have been making as to the alpaca, must be made in reference to the Chinese sheep, whose astonishing fecundity, it was thought, rendered them a desirable acquisition. *Ong-ti*, it seems, is a misnomer; the proper name is *Ti-yang*. Of this breed there are at Paris two varieties. The fecundity of that without ears, brought from London, has not come up to expectation. It first produced only one young one, then three lambs, and again two. The variety with ears has produced in the garden of the Imperial Society five lambs at a birth, of which only two have survived.

In Ireland these Chinese sheep are found not to answer. The wool is inferior, butchers will not buy the crosses, and the prolific energy is abated by crossing. In England also the last report is unfavourable, unless with reference to some cross-breeds. The Chinese have what we deem an excessive esteem for mutton. From 'Memoirs of the Jesuits at Pekin,' long ago, we learn that they thought this esteem well founded; and they record the fact that the soup and flesh of the sheep are believed to be good for invigorating old people, students, women worn out by child-bearing, and patients recovering from dysentery. They forbid the liver to be eaten even by the poorest, because of opinion that in a hundred sheep more than ninety have diseased livers. Believing in the doctrine of the transmission of hereditary qualities, it is not comfortable to apprehend an addition to the ills that afflict humanity from the introduction of earless or ear-bearing Ti-yang sheep, with livers in a state of dubious sanity. Upon the whole, with our present measure of light, we do not see our benefit in making very strenuous efforts to acclimatise among us these woolly natives of the Celestial Empire. We may well be content to wait the issue of the experiments in France—and this all the more readily when we find a member of the Imperial Society energetically protesting to his learned colleagues that their expectations in regard to the Ti-yang sheep are delusive.

Since 1854 France has succeeded in acclimatising another wool-producing animal, also to be classed among those the acquisition of which to our textile industry should be a matter of interest,—we mean the

z

goat of Angora, in Asiatic Turkey. In its habits it has so little of the vagrant propensities of the goat tribe that it is easily managed in flocks like sheep. Its introduction into France was owing to the impression made on M. de la Sagra, commissioner of the Government of France to the Great Exhibition of 1851. The beautiful tissues of Angora, especially the silk velvets, convinced him that the Angora goat ranked high among those wool or almost silk producing animals which ought to be introduced into France. Within sixteen months after the formation of the *Société d'Acclimatation*, a flock of nearly a hundred of these goats was landed in France. No time was lost in locating them amongst the Alps, the Jura, the mountains of Algeria, and more recently of Auvergne. The two places where they have best succeeded are the Alps of Dauphiné and Le Cantal. The fact of their permanent acquisition is regarded as settled. The vexing doubt arose, will the wool produced under such new circumstances preserve its silky texture? In the opinion of M. Davin, a great manufacturer and a most competent judge, the fleeces have in no degree degenerated. Of this the public had recent opportunities of judging in the Great Agricultural Exhibition in the Palace of Industry. From the flock at Cantal were taken the fleeces from which were manufactured those light silky stuffs, and those magnificent silk velvets, which were so greatly admired.

"If so much has been done in five years, where shall we be after ten?" asks M. Saint-Hilaire. "We do not know," he replies. "At least the most difficult part is done. It is a mere question of time; and, a little sooner or later, the Angora goat will definitively take its place, in our agriculture and our textile industry, among the merinoes, which France owes to Dabenton; the yak, which has just been given to her by M. de Montigny; and the alpaca, which will come to her in its turn."

These are great achievements of acclimatisation in France. We hope that this country has entered on the same career of patient study, with the view of ascertaining how the far-spreading dominions subject to British sway may be most speedily stocked with the various animals suited to their diverse climates. No nation in the world has such an interest as we have in such experiments; and yet we may be said to be only beginning them now. Though one of the primary objects of the Zoological Society of London was the introduction of exotic animals for ornament or use, this has been so far departed from that the well-stocked garden of the Society can only be looked upon as a collection of animals brought together for the illustration of the principal forms of animal life. The utilitarian aspect of zoology as the introducer of useful species of animals is prominently presented to our notice by the British Acclimatisation Society, founded in 1861, and having, we are glad to observe, a branch society in Scotland, with its headquarters in Glasgow.

The acquisition of new and ornamental poultry and game-birds, if successfully carried out, is sure to excite public interest; and we shall watch with solicitude for further reports on the introduction of the Honduras turkey, Chinese land-grouse, Canadian grouse, prairie grouse, American quail, &c.

The piscicultural operations of the Parisian Society are so varied and important that we do not attempt a description of them at present. We content ourselves with pointing out, with the help of information furnished by our French friends, that the Acclimatisation Society of Great Britain has committed a grievous mistake in regard to what our newspapers describe as "the new fish." In its First Annual Report (1861) we read, "We want a new pond-fish. Your secretary has been enabled to enlist the co-operation of Dr Gunther of the British Museum, a gentleman whose extensive knowledge of fish has gained him a European fame as an ichthyologist: this gentleman has given his verdict against *Lucio perca*, and this shows the value of not being in a hurry in choosing objects for acclimatisation; but he has, instead, highly recommended two fish, viz., the *Silurus glanis*, and also the *Guaramier* (*Osphromenus olfax*), which is pronounced to be the very best fish in the world. This fish is a native of Jamaica, but has been taken alive to the Mauritius."

So far as the Counani—or, as the French call it, Goorami—is concerned, we heartily say, Amen! seeing that the testimony as to its delicious qualities is unanimous; and we rejoice to know that last spring thirty specimens have reached France. The description of its edible qualities is quite appetising; in Guiana, whence we can procure it, it is called the sick man's fish. But as to introducing into our rivers or ponds the *Silurus glanis*, or European glanis, we say, Forbid it, *Pisces!* It is an ugly, voracious, mud-haunting creature, agreeing with the, we hope, vigorous digestion of Dr Gunther.

As for ourselves, we have no desire to taste a fish which is so voracious that it has been known in several instances to devour children. We happen to be fond of children, and to have half-a-dozen whose paternity is duly ascribed to us in the parish register, and who, moreover, are much addicted to dabbling in the lake in front of our dwelling: we are therefore disquieted by the thought that the said lake may be deemed a fitting habitat for the Silurus, the introduction of which may wring our heartstrings by the ugly monster, six feet long, it may be, and weighing from two to three hundred pounds, gobbling up our youngest-born. And this accompanied by digestive remorse, should we chance to eat of this destroyer of our family peace! for, according to Mr Yarrell, its flesh, being luscious, soft, and difficult to digest, is not suited to weak stomachs. And yet we are required to congratulate the British public because the London Society of Acclimatisation has succeeded in introducing into its ponds at Twickenham fourteen specimens of this ugly, eel-like, voracious fish. We devoutly hope every one of them may perish,—a hope which is not unfounded, seeing that all specimens

The Silurus Glanis.

of the Silurus introduced into France have perished. In the name of all that is reasonable, why think of introducing into our lakes or rivers a monster fish, with which our noble salmon will have no chance, and which will, doubtless, devour our best fish of all sorts by the hundredweight, seeing that it has a gullet big enough to admit a child of seven years old? We trust the Thames will be on fire with wrath at this ill-judged proceeding of the English acclimatisers, who from their French allies might have derived information which must have deterred them from acting on the advice of Dr Gunther. We hope they will peruse *Proces verbaux* in the bulletin of the Imperial Society for last January. M. Millet is of opinion, justified by various facts adduced, that the introduction of the Silurus into the rivers of Britain, abounding with salmon and valuable fish of various kinds, would be a deplorable mistake; that on the one hand, on account of its voracity, it will absorb a large quantity of excellent flesh, and give in return what is inferior in quantity, and especially in quality; and that, on the other hand, it will speedily prove a serious obstacle to the increase and propagation of good kinds of fish. The *lota*, which may be termed a miniature Silurus, having been introduced into the lake of Geneva, has bred so abundantly as to be reckoned the chief cause of the disappearance of the excellent trout called "La ferelle du Léman." The attention of the English Society could not therefore be too soon called to the inconvenience and danger of introducing the Silurus into the waters of England, especially those having trout.

M. Quatrefages corroborated M. Millet, adding that he had several times tasted the Silurus, and that he found its flesh neither delicate nor agreeable. M. Martin de Moussy assented, and stated that the Silurus abounds at the mouths of the rivers of America, and that its flesh is in such little esteem as only to be eaten by the most miserable of the population.

Mr Buckland, writing on 21st August 1865, seeks to tranquillise us. "Do not fear, we are not going to introduce the Silurus into the Thames. It is a pond-fish, and will only be placed in ponds. He is much too valuable to be placed in the Thames. There are plenty of ponds about full of wretched roach and dace, that breed naturally to an enormous extent. Master Silurus is just the boy to keep these fellows in order. Besides, please recollect that in his native home he lives principally upon *frogs*."

Both in England and France persevering efforts are being made to introduce the ailanthus silk-worm (*Bombyx cynthia*). Among the successful experiments is that of Lady Dorothy Neville at Dangstein, near Petersfield, Hants. Mr Buckland thus describes his visit to her plantation :—

"Her ladyship has set apart a portion of her beautiful and well-ordered garden, and has planted it with the young ailanthus trees, covering them over with a light canvass-made building—a precaution rendered necessary by the birds, which pick off the young worms. On entering this building I saw for the first time the living worms ; they were in the highest state of perfection, and really beautiful things to look at ; not white-faced pale-looking things, like the common silk-worm, but magnificent fellows, from two and a half to three inches long, of an intense emerald-green colour, with the tubercles tipped with a gorgeous marine-blue. Her ladyship pointed out to me how the silk-worms held on to the leaves ; they cared nothing for rain, less for the wind. Their feet have greater adhesive powers than the suckers of the cuttle-fish, and their bodies are covered with a fine down, which turns the rain-drops, like the tiny hairs on the leaf of the cabbage. Lady Dorothy Neville explained how readily, and at what little expense, they were cultivated, and that she had found a ready market for all the cocoons she could grow—a gentleman in Paris having offered to take all she could supply for French manufactures."

Her ladyship herself gives the following report of her experiments :—

"Of the silk-worms I have nothing at present to say, as they are not yet come out; but last summer I netted over three dozen trees, and placed 500 worms on them. They yielded 480 cocoons. A bird got under the nets, and took off some before it was arrested in its mischievous career. No wind or weather seemed to hurt the worms, and we kept some of the later ones on the trees when even the leaves were frost-bitten, but the worms did not seem to suffer. I have no doubt as to their hardiness. The three dozen trees would have fed at least 2000 worms, if we had had them, as the more the worms devour the leaves the stronger the latter shoot forth."

These facts seem to establish not only the probability of cultivating the ailanthus silk-worm in this country, but the ease with which it may be carried out. The shrub itself grows hardily and abundantly in the country already, and it may be seen flourishing in several of the gardens of the metropolitan squares, notably Belgrave and St James's. It will, indeed, live anywhere, and delights in poor and sterile soil; and where it lives the worms will live also.

"The Council cannot but think that the general introduction of this new form of cultivation would be most beneficial, as it could be carried out on any even the smallest scale by every cottager or small landowner who has a garden; and a ready market may be found for the smallest parcel of cocoons. To ladies especially this operation may be recommended by the fact that they may, without the slightest hyperbole, grow their own silk dresses in their own gardens."

These are hints worthy of the attention of many who complain that they "have nothing to do," as well as of those who "have nothing to wear;" and for their information we add that Lady Dorothy Neville has published a pamphlet on the culture of the silk-worm, in which they will learn how to kill *ennui* and clothe themselves in silk.

The following remarks, lately addressed to Dr Hoskyns by a gentleman at Salisbury deeply conversant with silk cultivation and manufacture, are very interesting and suggestive :—

"I have introduced a notice of the *Bombyx cynthia* as a silk-producing insect in some lectures recently delivered by me, and I am convinced that this branch of industry may be most profitably introduced into our union workhouses. There is a large amount of labour wasted, simply because it has not been profitably applied. Plant, therefore, the ailanthus shrub, and let the women and children attend to the worms. Pay them a percentage upon the result, and divide the inmates into sections, so that there may be honest rivalry. The sections would be stimulated to exertion by the personal interest each individual would have in the result, and section would soon compete with section for superiority. I see no reason why, in reformatories, penitentiaries, and the like, some effort should not be made to rear these worms. In fact, wherever there is unapplied child or female labour it can be advantageously introduced. The ratepayers would not alone benefit; habits of industry and method would be insensibly taught, and with care the present pauper might become a silk-grower either for the capitalist or on his own account."

We have not yet commended the doings of the acclimatisation societies to the support of British agriculturists, on the ground that they greatly concern themselves in adding to the number of vegetable substances fit for human food. The Bulletin of the French Society is full of interesting notices of plants and tubers, the acclimatisation of which is desirable. In the reports of the British Society we have an account of experiments on the Chinese yam, the introduction of which into France is already accomplished. All the accounts from those who have cultivated it in Guernsey are unfavourable, with the exception of that of Mr Carré, a gentleman who has paid great attention to the subject of acclimatisation. He recommends it as an article of food, and purposes extending his cultivation of it. His success leads to the conclusion that the failures in other cases are due to improper management. The roots have been tried both plain boiled and with sauce, and pronounced delicious. In texture and flavour they are excellent, and, if one vegetable can be compared with another, may be said to resemble very good mashed potatoes. The evidence of gardeners and others acquainted with the plant leads to the belief that the Chinese yam is an excellent vegetable, and, with proper cultivation, will grow to an enormous size. It should be planted in trenches, and the lighter and the more sandy the soil, the better it will thrive.

We are agreeably surprised to learn that the Brazilian arrowroot is very successfully grown in Guernsey. The produce of this plant is sometimes enormous. "From 1½ perch," reports Mr Martin, sheriff of Guernsey, who has zealously promoted the cultivation of this valuable vegetable, "I manufactured one year 60 lb. of arrowroot, which I sold at the rate of 1s. per lb., being at the rate of £193 the acre! I have never succeeded so well since. It has never failed, however, to pay me well for the ground it occupied and the labour required by it."

In the hope of inducing graziers to patronise acclimatisation societies, we must not omit to notice that the English Society is experimenting on bunch-grass, which seems likely to be of service on waste and common lands. In his 'City of the Saints,' Captain Burton gives this description of it:—"The Festuca is a real boon to the land, which without it could hardly be traversed by cattle. It grows in clumps upon the most unlikely ground—the thirsty sand and the stony hills;

in fact, it thrives upon the poorest soil. In autumn, about September, when all other grasses turn to hay, and their nutriment is washed out by the autumnal rains, the bunch-grass, after shedding its seed, begins to put forth a green shoot within an apparently withered sheath. It remains juicy and nutritious, like winter wheat in April, under snow; and, contrary to the rule of the Gramineæ, it pays the debt of nature, drying and dying about May: yet even when in its corpse-like state—a light yellow straw—it contains abundant and highly-flavoured nutriment. I brought back with me a small packet of the bunch-grass seed, in the hope that it may be acclimatised. The sandy lands about Aldershot, for instance, would be admirably fitted for its growth."

Enough has been said to demonstrate that they who expect great things from the utilitarian turn recently given to natural history are not indulging in chimerical hopes. Those of this generation have already witnessed greater accessions to the number of domesticated animals than were made for many previous centuries; and as it is by means of acclimatisation and hybridisation, or what Mr Darwin terms natural selection, that the earth has been furnished with all varieties of animals and plants, we are imitating nature when by science we bring about those modifications of animal and vegetable life which she effects slowly, because Providence wills that the greatest transmutations shall result from man's intelligent intervention. Nature does not produce the Montreuil peach, or the Fontainebleau grape. Wheat is nowhere found growing wild; and, as Buffon asks, to have modified a species of grass into wheat, is not this a kind of creation? Nature has not given us the Durham ox, the Southdown sheep, or the English race-horse. These—and a thousand other instances might be mentioned—are examples of the manner in which man, for his convenience and comfort, modifies the handiwork of the Supreme. We are far as yet from having reached the limits of our delegated sovereignty over all the things which have been "put under our feet." We have still much to learn as to how the earth has been peopled with all the species of animals and plants, and as to how far these may be introduced into any particular locality.

THE END.

www.ingramcontent.com/pod-product-compliance
Lightning Source LLC
Chambersburg PA
CBHW031421230426
43668CB00007B/392